THE ELUSIVE WOW

SEARCHING FOR
EXTRATERRESTRIAL INTELLIGENCE

ROBERT H. GRAY

PALMER SQUARE PRESS
CHICAGO

Publisher's Cataloging-in-Publication Data

Gray, Robert H.
  The elusive wow : searching for extraterrestrial intelligence / Robert H. Gray.
  p. cm.
  Includes bibliographical references and index.
  ISBN: 978-0-9839584-4-4
  1. Life on other planets. 2. Radio astronomy. 3. Cosmology—Popular works.
  I. Title.
QB54 .G739 2011
999—dc22

                                                                    2011915228

Jacket, case, and interior design: Kim Bartko, Bartko Design, Inc.

Cover: Globular cluster Omega Centauri. Image credit: NASA, ESA, and the
Hubble Heritage Team (STScI/AURA). Acknowledgment: A. Cool (San Francisco
State University) and J. Anderson (STScI).

Anonymous engraving from Camille Flammarion's 1888 book *L'atmosphère: météorologie populaire*. Source: Wikipedia.

CONTENTS

PREFACE

The "Wow!" signal was detected in 1977 during Ohio State University's long-running search for radio signals from the stars, and is the best candidate for a signal from extraterrestrial intelligence ever reported. I read about it several years later and called the Ohio State researchers to learn more. I was surprised to find that no other astronomers had followed up on the intriguing signal, since if confirmed it could lead to discovering life on another world, but things that go bump in the night without appearing repeatedly are easy to dismiss as local interference—and they usually are. Few astronomers elsewhere had the right gear to search for the signal even if they had been interested in doing so.

I visited the giant Ohio radio telescope and learned about its search for radio signals from the stars—the only full-time search anywhere at that time. The evidence for the Wow being a signal from a celestial source seemed very good. The rise and fall of the signal's strength during 72 seconds looked almost exactly like the signature of the antenna's beam sweeping over a celestial radio source, rather than local interference, and it seemed unlikely that an aircraft or satellite would mimic that special signature. But unlike natural radio sources that hiss across wide bands of the radio spectrum, the signal was narrow like radio signals usually are, concentrated in one channel out of the 50 frequency channels being monitored. And, unlike most natural sources, it seemed intermittent—there for a minute, then gone—not seen when a second beam swept past the same spot a few minutes later.

The Wow is intriguing because it had many characteristics expected of a radio signal from the stars, and it was the best candidate that the Ohio State search found during their long-running search covering more than half of the sky. It seemed possible that the Ohio scientists had glimpsed a broadcast from another world, but could not confirm it because their telescope could view that spot in the sky for only a few minutes each day.

I decided to hunt the Wow, despite the awkward fact that I'm not an astronomer. Since then, I've searched for it with a small home-brewed radio telescope, with the sophisticated Harvard-Smithsonian META radio telescope, with the Very Large Array—one of the world's most powerful radio eyes on the universe—and with the Hobart radio telescope in Tasmania in the Southern Hemisphere. The signal *has* proven elusive...

It's possible that it was just interference from a man-made gizmo like a satellite, but there are many reasons to think that it may have been a real signal from the stars, perhaps something like a lighthouse that shines our way periodically. All efforts to find it so far amount to only about one day of listening at any one spot, so it's possible an intermittent signal has been missed.

This book tells the tale of the elusive Wow and my search for it, and it also tells the much broader story of the search for extraterrestrial intelligence, known as SETI.

Part One of the book focuses on the Wow signal—the Ohio search that found it, what makes it a good candidate, and my quests to confirm it. Hunting the Wow will take us on tours of radio observatories, explain how radio telescopes work, show examples of what searchers see with interstellar radio receivers, and demonstrate how one goes about trying to solve scientific mysteries. I aim to give readers a sense of how it feels to hunt for broadcasts from other worlds, showing lots of illustrations; it might look a bit like a textbook at times, but only the most interesting parts are included. There's no happy ending to the Wow story, yet, but this is how someone might one day find ET.

I don't claim that aliens have been discovered, but I do claim that the Wow was a pretty strong tug on the cosmic fishing line. It's possible, however, that it's just a big fish story, a big one that got away, a red herring—so a brief book about SETI is included at absolutely no extra charge. Some people might want to read the second part first for background on astronomy and on the rationale for searching for signals from other worlds.

Part Two presents the reasons for thinking that planets, life and broadcasters might exist elsewhere, describes strategies used to search for signs of their presence and milestones in the history of searching, speculates on what we might find, and gives reasons to worry that searching could prove fruitless. The idea of looking for intelligent life elsewhere was pretty speculative before the recent discovery of extrasolar planets. Until then, it was not certain that any other stars *had* planets for life to live on. Now we know that planets orbit many other stars and probably zillions, so life has many possible homes. Whether any other planets actually have life is still a big question; many scientists think it's likely, and robotic explorers in our own solar system could soon send back a picture of a fossil or other evidence that says "Yes!" If life does exist elsewhere, it seems reasonable to think that intelligence might sometimes evolve, and to think that some smart critters would use radio or other technology that we might detect from afar. SETI lets us explore other worlds for intelligent life by listening for signals from the comfort of our own good Earth.

Unlike many writers on these topics, I'm not an astronomer or a journalist. I was trained as an urban planner and analyze data for a living—financial, economic, social, and scientific. I've learned enough about astronomy and radio to hunt for broadcasts from the stars, and I'll explain what general readers need to know as we go along. We'll dive deep into a few topics (the most interesting ones) but cover most pretty quickly. We'll also get a few grins poking fun at quirks of science culture, such as awkward acronyms, obscure jargon, and other easy targets.

Some people have the impression that a massive search for ET has been under way for a long time—that *that* is what all those big dish antennas *do*—but not so. During the last few decades, only one or two big antennas have been searching, typically seeing a spot of sky no bigger than the Moon at any moment and not seeing the other 99.999%. Most searches listened on thin slivers of radio spectrum for a few minutes, and just a few have peeked at the optical spectrum. The sky could be blaring with radio super-stations whose frequency we've not yet tuned, or twinkling with laser beams we've not yet looked carefully enough to spot.

Searching for other beings on other worlds is an exciting opportunity of our time. There may be many New Worlds waiting to be discovered by intercepting radio or light waves rippling through space, just as there were new worlds to be discovered on our own planet by sailing the seas in earlier eras. Signals launched long ago might be raining down on the landscape around us right now, or could appear tomorrow in a previously quiet patch of sky. All we need to do is look up with the right instruments and we may find that some of the myriad stars are Suns to someone or something—and that some of those others might have interesting stories to tell.

ACKNOWLEDGMENTS

I'm deeply indebted to Bob Dixon, Jerry Ehman, and John Kraus at Ohio State University for sharing information about the "Wow!" signal.

I'm especially grateful to the people who helped me use big radio telescopes to search for the elusive Wow: Paul Horowitz for trusting me with the Harvard-Smithsonian META observatory on several occasions, and Patrick Palmer and Kevin Marvel for helping me get access to and learn how to operate the Very Large Array. I'm sure they were all as relieved as I was that I didn't break anything! Simon Ellingsen at the Mt. Pleasant observatory in Tasmania did all of the observing in our collaboration, and I got the easy part analyzing data back in the USA but regret missing a trip down under.

Other very notable help came from Skip Crilly, then at Hewlett-Packard, who custom-built a high-tech microwave radio receiver, and Barney Oliver and Charles Kingsford-Smith, both then at HP, who arranged for the donation of a digital spectrum analyzer when such things were very new and awfully expensive. William "Tap" Lum at the University of California at Berkeley donated a low-noise amplifier he'd built that was better than ready money could buy, and Bill Jayne and Terry Lilley built custom computers. John Pattee of Pioneer Hill Software provided several generations of his sophisticated SpectraPlus spectrum analyzer software. The SETI Institute and the Planetary Society helped defray costs of several of my campaigns to find the elusive Wow. Ted Molczan performed orbital calculations showing that the previously classified U.S. reconnaissance satellite programs Big Bird and Key Hole are unlikely to account for the Wow signal, because the satellites were not in the vicinity at the time it was recorded.

People who commented on drafts of this book include Marc Abel, Peter Backus, Bob Dixon, Jerry Ehman, John Kraus, Kevin Larmee, Steve Lord, Jon Mills, Patrick Palmer, and Rory Sellers.

Many people and organizations provided permission to use photographs and illustrations, especially Seth Shostak of the SETI Institute and the SETI Institute itself. The translation from the philosopher Epicurus by Cyril Bailey is used by permission of Oxford University Press, and the translation from the poet Lucretius by Professor Mary-Kay Gamel is used with her permission.

Kim Bartko of Bartko Design designed the cover and the interior of the book. Copy was edited by Joanne Asala of Compass Rose Horizons. Amir Alexander used the phrase *Elusive Wow* in the title of a magazine article he wrote on the topic, which helped convince me it was the right title for this book.

My wife Sharon Hoogstraten was as patient as a saint, our children Alexander and Dillon were only slightly skeptical, and my parents Marian and Paul were surprisingly tolerant of a son looking for aliens.

PART I

The Elusive Wow

PROLOGUE

One summer night in 1977, a huge radio telescope in the countryside near Delaware, Ohio, began receiving a radio signal from the sky—a signal that might have been broadcast from another world. Nobody was present in the underground control room at the time, and there was nothing to hear, anyway; the signal was just a rising voltage in one channel of a receiver chilled in liquid nitrogen and monitored by a computer along with 49 other channels tuned to nearby frequencies. Every 12 seconds, a computer printer hammered out a line of numbers keeping a crude record of the radio rain sprinkling down on the big antenna from the sky.

The signal strength increased for a half-minute, as celestial radio sources always did when the antenna's searchlight beam began sweeping across them. Running out of single-digit numbers to record the surging strength, the computer began printing letters of the alphabet—"A" for ten times higher than the background noise level, B for 11 times higher, all the way up to U for 30 times—much stronger than any natural radio source in that part of the sky. Then the signal slowly faded as the antenna's beam swept past the source, disappearing back into the background static.

Several days later, project scientist Jerry Ehman began flipping through hundreds of pages of accumulated printout looking for the signature of a radio signal from another world, as he had done many times before. He saw the pattern he was seeking after just a few pages and scribbled "Wow!" in the margin, naming one of the best candidate interstellar signals ever seen.

The Ohio scientists looked for the signal on roughly a hundred subsequent days without finding it again, but their antenna could see that spot in the sky for only a few minutes each day so their efforts totaled only about four hours. John Kraus, the observatory director, published an account of the event in a scientific magazine of modest circulation, and the telescope returned to its survey for radio broadcasts from the stars.

CHAPTER 1

THE "WOW!" SIGNAL

BIG EAR

A huge antenna was erected in the countryside near Delaware, Ohio, between 1956 and 1963, built with the ambition of being one of the world's premier radio telescopes. These novel gizmos were revealing an alien sky where stars are dark but previously invisible objects are bright, some blazing furiously spewing out radio waves—a celestial zoo of pulsars, quasars, black holes, and other bizarre objects. Early radio telescopes like this one were basically just radio receivers with big antennas pointed skyward, and celestial objects shining at radio frequencies made a hiss in the receiver when the antenna was pointed toward them.

The antenna's designer was John Kraus, a physicist and electrical engineer at Ohio State University in nearby Columbus, who was playing a major role in the new science of radio astronomy. Kraus was a master of antenna design—the inventor of the helical antennas that sprout from many spacecraft—and had conceived a "Big Ear" that would intercept more radio sizzle from distant objects than any other existing radio telescope. It would be too big to track

celestial objects by moving; it would instead passively survey the sky by letting the Earth's rotation sweep its searchlight-like beam across the heavens.
Construction was slow, modestly funded by the National Science Foundation,
with students working as engineers, electricians, and steelworkers.

In radio astronomy, you study the radio glow from distant objects to learn
something about them, just as you study light in the more familiar optical flavor of astronomy; radio and light are both electromagnetic radiation. To collect the radio rain sprinkling down from the sky, you need big antennas—big
buckets, in a sense. A parabolic dish-like shape has just the right geometry to
reflect waves hitting it from directly in front so that they converge at a single
focal point, concentrated and all arriving at the same instant, where you put a
detector of some sort. Most optical telescopes use a mirror with the magical
parabolic shape to do the reflecting, while "dish antennas" use a metal surface
or mesh with the parabolic shape for the purpose. The reflected waves generate
small voltages in the *real* antenna located at the focus of the dish, often just a
stubby little metal rod. After lots of amplification, the voltage is measured to
find the radio brightness in the direction the reflector is pointed.

Figure 1.1. The Ohio State University Radio Observatory. The flat structure at right was tilted to
reflect radio waves from the sky into the fixed parabolic section on the left, which reflected them
back to converge into funnel-like feed horns on the ground in front of the flat reflector. Waveguides
carried signals to an underground receiver room. The area between the reflectors was the size of
three football fields. Photo courtesy of North American Astrophysical Observatory.

The good news is that radio telescopes can see a previously invisible universe—stuff glowing at radio instead of optical wavelengths. The bad news is that much bigger telescopes are needed than with light, because radio waves are much longer than light waves—roughly a million times longer at the frequencies used in much of radio astronomy—and longer waves need wider reflectors. To get the same spatial resolution or sharpness of view as even a small optical telescope would require a dish many miles across, but it's very hard to build them much bigger than about 300 feet because gravity makes them sag out of shape when pointed in various directions.

Kraus's strategy for making a really big antenna was to build a slice through a 360-foot dish—a strip only 70 feet high but the full 360 feet wide—with the bottom edge supported by the ground so it would not sag. Anchored to the ground, it could not move to point at objects in the sky, so he bounced radio waves into it using flat panels 500 feet away, tilting them to point upward at various angles and letting the Earth's rotation sweep the antenna's view across the sky. Figure 1.1 shows the finished antenna.

When it began operating in 1963, the telescope delivered a remarkably detailed view of the radio sky, as shown in Figure 1.2. Its high sensitivity came from its large surface area, equivalent to a dish 175 feet in diameter. Its spatial resolution was as good as a 360-foot dish in the horizontal dimension

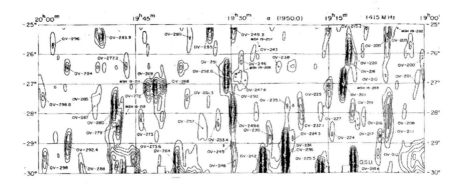

Figure 1.2. A radio map from the Ohio Survey at 1415 MHz covering the area where the "Wow" signal was later detected, although no sign of it is visible here. Source strength is shown by contours, and sources appear elongated because of the shape of the antenna, but most are actually point-like. The horizontal axis is right ascension, and the vertical axis is declination. Courtesy of North American Astrophysical Observatory.

(right ascension, in astronomer's jargon, like longitude projected on the sky) but only one-fifth as good in the vertical dimension (called declination, like latitude) because it was only 70 feet high. Big antennas are directional like searchlight beams, and bigger antennas have narrower beams. Ohio State's beamwidth was about two-thirds of a degree tall and one-eighth of a degree wide, a spot on the sky a bit taller than the Moon (which is a half-degree across) and a quarter as wide.

The observatory surveyed the sky for years and discovered thousands of new radio sources glowing at a frequency of 1415 MHz (frequency is measured in cycles per second but is termed Hertz, after Heinrich Hertz who first demonstrated radio; a frequency of one million cycles per second is termed one megahertz and abbreviated MHz). The Ohio Sky Survey produced maps and catalogs with nearly 20,000 radio sources, more than half of them previously unknown; some turned out to be among the most powerful and distant objects ever found.

By 1972, the big survey was finished. Larger telescopes had been built elsewhere, some using the Kraus design, and the interests of funding agencies were changing. One of the biggest radio telescopes on the planet was left with no financial fuel for astronomical research.

THE FIRST FULL-TIME SEARCH FOR ET

Bob Dixon, a former student and later collaborator of Kraus's, spent the summer of 1971 at NASA's Ames Research Center in Mountain View, California, where several dozen scientists and engineers gathered to brainstorm about detecting radio signals from other worlds. Experts in astronomy, biology, computers, design of antennas, engineering, and many other areas developed preliminary designs and published their findings as *Project Cyclops: A Design Study of a System for Detecting Extraterrestrial Intelligent Life*. The book argued that intelligent beings might exist on planets orbiting some other stars, and that we could detect their broadcasts with special receivers which were described in detail. The report proposed a search that might have required hundreds or thousands of big antennas and cost billions of dollars. That ambitious Cyclops system was never built, but the study blazed a trail for later searches and laid the foundation for a NASA search that would go on the air nearly 25 years later.

When Dixon returned to Ohio, he proposed using the unemployed telescope to begin the world's first full-time search for interstellar broadcasts, and Kraus agreed. They lacked funding but had a huge antenna and a good receiver; by December of 1973, they were on the air listening for signals from other worlds. They used an eight-channel receiver tuned near the 1420 MHz frequency of hydrogen thought to be attractive for interstellar broadcasting, a part of the radio spectrum also known as 21 centimeters because of its wavelength (scientists like to use frequency and wavelength interchangeably; waves arriving 1,420 million times per second and traveling at the 300,000 km/sec speed of light have peaks about 21 centimeters or 8 inches apart). The system ran night and day recording the strength of the receiver's output at each frequency on a long roll of strip chart paper, and Dixon and colleagues scanned miles of charts looking for the signature of a radio signal from the stars. They slowly surveyed the 70% of the sky visible from their site after the fashion of their earlier astronomical survey but now looking for radio transmissions.

For more than ten years, the Ohio State project was the world's only full-time search for ET (Project META at Harvard was the second, beginning in 1985). A dozen searches had been done elsewhere, but most were brief and covered only short lists of stars. Ohio was the first to survey a large part of the sky with something that could be fairly called an interstellar radio receiver.

By 1977, a 50-channel receiver was installed and the search was computerized. The "front end" of the receiver was a low-noise amplifier chilled by liquid nitrogen to make it as quiet and sensitive as possible. Each channel consisted of a 10 kHz-wide slice of the radio spectrum, like 50 AM radios tuned to adjacent frequencies. The receiver outputs were fed into an IBM 1130 computer—resembling an office desk more than a computer—which averaged the voltage of signals in each channel for ten seconds to increase sensitivity, like a long photographic exposure. After digesting this for two seconds, the computer would print a line of numbers to record the intensity of the noise detected in each of the channels, hammering out roughly a hundred pages of fan-fold paper each day.

To reduce interference from local transmitters, two feed horns were used to collect the radio waves focused by the big reflector. They were oriented side by side but slightly offset, so that a celestial source would be detected first in one and then in the other a few minutes later. The receiver subtracted whatever one horn was hearing from the other, so local signals appearing in both horns at the same time would cancel out, but signals from a celestial source would not cancel because they appeared in just one horn at a time. Two feed horns gave the antenna two beams, like two searchlights side by side, and a celestial source would normally be detected twice several minutes apart as each of the beams swept across it.

This was the Ohio group's interstellar radio at the time the Wow was detected—big, cleverly rigged to filter out interference, and with a record of having discovered half of the radio sources known.

THE WOW SIGNAL

On August 15, 1977, the antenna was pointed about 20 degrees above the southern horizon. As night fell, the beam swept across the Milky Way, lighting up many channels with the radio glow from clouds of hydrogen gas in the galactic plane, then the hiss in the receiver subsided back to the usual steady background level.

Just after 11:15, channel number two of the receiver began registering a signal. The signal's intensity increased during the next half minute as the most sensitive center of the beam swept closer to the source, rising to an unusually strong 30 times the background level when the antenna pointed straight at it. The intensity of the signal then decreased for the next half minute as the Earth's rotation swept the beam past the source. Nobody was there to hear or see it; the printer just hammered out a line of numbers every 12 seconds as it always did while recording the radio sky.

Jerry Ehman later reviewed the accumulated computer printout. He had worked on the big astronomical survey and was the project's volunteer computer wizard. He wrote parts of the computer program that looked for signals, cleverly working within the machine's absurdly tiny but fast 16-kilobyte memory. Ehman had a sharp eye for interference, having seen and rejected many false leads; he knew exactly what the signature of an interstellar radio transmission should look like.

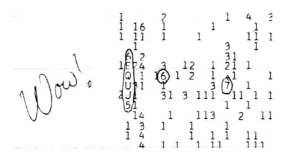

Figure 1.3. Detail of original printout showing Jerry Ehman's "Wow!" annotation and the signal recorded as the characters 6, E, Q, U, J, 5. Courtesy of Jerry Ehman and Bob Dixon, North American AstroPhysical Observatory.

When he saw the sequence of symbols representing the signal, he circled them and scribbled "Wow!" in the margin, because it was exactly what the search sought—a radio signal from the sky, seemingly from a celestial source. It was much stronger than the background noise and matched the signature of a celestial source passing through one of the antenna's beams. The absence of responses in adjacent channels was the signature of a radio signal, as opposed to a natural source which would light up many channels. What he saw is shown in Figure 1.3. A slightly edited version is shown in Figure 1.4, including all 50 channels and with the letter codes converted to numbers.

A chart makes it easier to visualize what the telescope saw. Figure 1.5 shows the intensity in each channel for the six time periods during which the

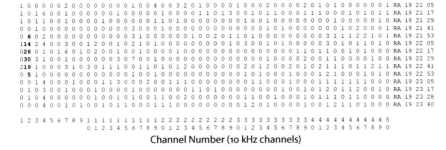

Figure 1.4. Intensity of the receiver responses for all 50 channels, recorded at 12-second intervals. The Wow signal is the vertical series of numbers 6, 14, 26, 30, 19, 5 in the second column from left, converted from the original printout which used letters to represent intensities over nine. The right ascension coordinates on the right include some but not all corrections. Data courtesy of North American AstroPhysical Observatory.

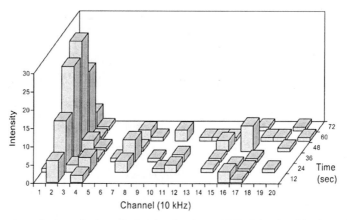

Figure 1.5. Intensity in each channel, showing the rise and fall of the signal in channel 2. Channels 21-50 contained no features stronger than 3 sigma and are not shown. Data courtesy of North American AstroPhysical Observatory.

signal was recorded. The rise and fall of intensity in channel 2 is due to the antenna's bell-shaped beam sweeping across the source; the signal's strength was probably constant.

The Wow was the strongest signal ever detected by the Ohio State search, other than obvious interference. Its intensity was recorded in units of sigma— how many times stronger it was than the average variation in the background noise (called standard deviation in statistics jargon; called root mean square or rms by engineers). Statistics says that in a system making 250 averaged measurements per minute like Ohio's (50 channels, five times a minute), random noise might produce a value as strong as three sigma every few minutes or four sigma every hour (going as $\log(n)$), but bigger peaks would be rare and a 30 sigma peak would not be expected in thousands of years.

The Ohio group knew that the Wow was not a natural radio source because their own earlier survey had found none in the vicinity. They found no record of spacecraft that might account for the signal, although secret satellites could not be entirely ruled out. But the frequency band where the signal was seen is reserved for radio astronomy under international agreements and is monitored by astronomers of many nations, which would make it an awfully conspicuous choice for clandestine satellites. Aircraft, near-Earth orbiting spacecraft, and other possible local sources of interference seemed unlikely since the signal passed through the beam in just the time expected of celestial sources.

One aspect of the event was puzzling: a celestial source was always detected twice—first in one beam, then in the other beam a few minutes later as the source passed through the side-by-side beams—but the Wow signal was detected in only one beam.

A simple explanation for the single-beam detection would be a source turning on or off during the few minutes between the two beams sweeping across it. Few natural radio sources appear and disappear in a matter of minutes, and almost all natural sources make a wideband hiss that would be spread over many channels. Radio transmissions, however, switch on and off easily and might appear intermittently. Many kinds of interstellar transmissions might be intermittent—a narrow beam pointed our way from time to time, or a broadcast in all directions turned on only now and then, or perhaps an antenna something like Ohio State's on a distant planet sweeping its beam across the sky shining our way periodically once each "day".

Alternatively, the signal might have been changing in frequency in just the right way to have drifted into or out of the band during the time between beams sweeping past the source. That would require a fortuitous draft rate, but with the signal detected in channel 2 near the edge of the band, it's a possibility.

The Ohio search had found just what it sought—a radio signal that appeared to come from a celestial source—a candidate interstellar broadcast. But the signal would have to be detected repeatedly to confirm that intriguing possibility. Ohio looked on approximately one hundred days during the next few years, although for only a few minutes each day and totaling about four hours. Finding nothing, they resumed their all-sky survey. Their system was best at mapping the sky, discovering sources that others could study by tracking them with steerable telescopes.

CHAPTER 2

A CANDIDATE INTERSTELLAR SIGNAL

What does it take to make a good candidate for a signal from another world?

For starters, the signal must come from the sky, from a position fixed amid the stars. And it should be clearly artificial—more like a pure tone than the noisy hiss of natural sources, or different in some other way. An artificial signal that seemed fixed amid the stars is just what the Wow! appeared to be. This chapter shows why the signal seems like a good candidate, even though seen only once. The Wow is like a tug on a cosmic fishing line; it felt like something nibbling, although there's no proof that we got a call from ET.

CELESTIAL SIGNATURE

There are several reasons to think that the Wow came from a celestial source rather than a local transmitter. First, its strength showed the rise and fall always seen when celestial radio sources passed through the antenna's beam.

Second, the time it took to pass through the beam was almost exactly right for a celestial source viewed from the turning Earth with that particular antenna and in the direction it was pointed. Third, there was no sign that the source was moving like an aircraft or spacecraft.

The smooth rise and fall of intensity is the signature of a big antenna like Ohio State's sweeping across a radio source. The hiss of the receiver rises in strength as the beam approaches a source, is loudest when the antenna points straight at it, then fades away as the Earth's rotation sweeps the beam past— rather like the intensity of sound as a vehicle approaches, passes by, and then recedes into the distance. Figure 2.1 shows that the six intensities recorded for the Wow fit the bell-shaped antenna pattern almost perfectly, a sure sign that the signal came in through the antenna's skyward-pointed beam as the Earth turned rather than from a ground-based transmitter.

The amount of time it took the Wow to pass through the antenna's beam very nearly matches the expected transit time for celestial sources in the direction the antenna was pointed, which is another sign that the signal came in through the antenna. Sources should take about 36 seconds to transit the most sensitive middle half of the beam called the half-power beam width (HPBW

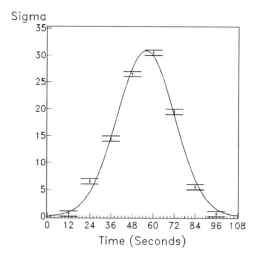

Figure 2.1. The intensity of the Wow signal in channel 2 and a fitted curve. The intensity measured at 12-second intervals is shown by small crosses, while the line is the Gaussian curve fit by statistical procedures and is the shape expected of the antenna's beam pattern. Error bars show the uncertainty of measurements in time (±6 seconds along the horizontal axis) and intensity (±0.5 sigma along the vertical axis). Data courtesy of North American AstroPhysical Observatory.

in unpronounceable jargon, the part halfway up the curve), and the signal took about 38 seconds—pretty close to the expected time, especially since the beam was known to have broadened a bit as the antenna aged. Transit time could range from just 32 seconds with the antenna pointed toward the celestial equator (declination 0°) to forever if pointed at the celestial north pole (declination +90°, essentially the north star). The Wow's transit time was close to matching even this detail, good evidence for celestial origin.

An aircraft or nearby satellite would zip through the beam much quicker. Visualizing the beam projected on the sky, its width was about one-quarter of the Moon and its height a bit taller; an aircraft would cross such a small angle quite quickly. Satellites in low-Earth orbit shoot across the sky at four degrees per minute, crossing the beam in just a few seconds.

Secret satellites were one possibility that Ohio could not rule out, since the information was not available, and some people have suggested that the Wow signal may have been a transmission from a spy satellite. Orbits of some U.S. reconnaissance satellites have since been made public, and we now know that none of them were crossing Ohio's beams at the time the Wow was detected. There could be still other spy satellites, but that origin now seems less likely. John Kraus calculated that a spacecraft would need to nearly as far away as the Moon to mimic the transit of a celestial source, and few if any satellites are at that range.

Having the same antenna signature as a natural celestial source, you might think the Wow was one. But there aren't any there.

NO STRONG NATURAL SOURCES

It's very unlikely that the Wow was a natural radio source because the only ones in the vicinity are weak, about one-thousandth of its strength.

The strength of the signal is thought to have been in the range of 54 to 212 janskys (according to estimates by Jerry Ehman and Russ Childers, respectively), which is quite strong as radio sources go. Natural sources range from a few thousand janskys for the strongest sources other than the Sun, down to thousands or millionths. The jansky is a measure of spectral power named for Karl Jansky—the amount of power that falls on a given area and over a given amount of the spectrum, and one jansky is 10^{-26} of a watt falling on an area of one square meter and in a one Hertz slice of frequency, written as 1 x 10^{-26} W/m²/Hz or 1 Jy. For those unfamiliar with this scientific notation, it replaces insanely awkward numbers like 0.00000000000000000000000001 with compact ones like 1.0×10^{-26}, where the negative exponent tells you how many places to move the decimal point to the left; a positive exponent means you move the decimal point to the right, so 1×10^2 or 10^2 means 100.

Ohio State's earlier astronomical survey found no sources near the Wow locales down to one-tenth of a jansky, a sensitivity where more than ten thousand radio sources can be detected. More sensitive later observations found a few hundredth-jansky sources, but at that faint level roughly a million sources glow across the sky, and some would fall near the Wow locales by chance.

NO SUSPECT STAR

Stars don't shine at the hydrogen frequency (only *cold* hydrogen does), so finding a prominent star at the Wow locales would be surprising—hinting that the source of the signal might be somebody living in that neighborhood.

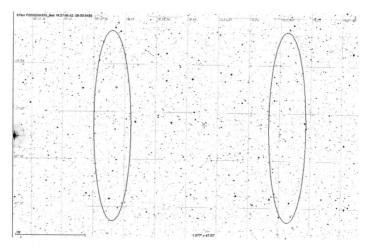

Figure 2.2. Stars in and around the antenna's two beams at the time of the Wow peak intensity. The horizontal axis is right ascension and the vertical axis is declination; the source should be near the vertical axis of one of the two beams, at right ascension 19ʰ 22ᵐ 22ˢ (right) or at 19ʰ 25ᵐ 12ˢ (left) both at declination -27° 03′ (epoch 1950). Credit: STScI POSS2UKSTU_Red visualized with Aladin.

Unfortunately, the beams covered many stars, shown in Figure 2.2, and none of them seem likely suspects. A check of three sun-like stars in the vicinity was done in the early 1980s by David Talent, who examined their light for unusual features such as spectral signs of artificial elements that a civilization might spice its star with to mark it as the home of intelligence, or perhaps dumping radioactive waste. No strange spectral features were seen, but lots of other stars were in the beams and have not been checked.

NARROW BANDWIDTH

The antenna said the signal had the signature of a source in the sky, but the receiver said it was a radio transmission. Most natural sources are noise-like and spread over a wide band of frequencies, but the Wow signal did not spill out of a 10 kHz-wide channel—the width of an AM radio station—which is strong evidence that it was a radio signal.

Natural radio sources come in two flavors. Continuum sources radiate across much of the spectrum, and would light up many or all channels of the receiver, while spectral line sources radiate at a single frequency, although the frequency is almost always broadened by random motions in the clouds of interstellar gas where the lines originate. No way is known for either sort of source to produce an emission as narrow as the Wow. Its frequency was near the 21-cm hydrogen spectral line, but hydrogen is always broadened to roughly 100 kHz or 10 Ohio channels, and its intensity has never been seen to flare up or disappear over a few minutes.

Radio signals, on the other hand, are usually narrow. Transmitters concentrate their energy in a narrow range of frequency, and receivers use a tunable filter to pass just the slice of frequencies containing a desired signal. That lets a signal of interest stand above the clamor of noise that fills the electromagnetic spectrum, such as unwanted chatter at other frequencies, natural snap crackle pops from lightning and other causes, and hiss in receiver circuits due to electrons flowing through them. The fact that the Wow did not spill out of a single channel is a strong sign that it was a radio signal, although it says nothing about where it came from; many man-made radio signals would be just as narrow.

A MAGIC FREQUENCY?

Ohio State was listening near the hydrogen frequency because some scientists had suggested it as a likely choice for first contact between stars—broadcasters might choose a "magic" frequency that searchers could guess, without tuning across the vast radio spectrum.

The Wow frequency was reported as 1420.3556±0.005 MHz, which is within about one 10-kHz channel of where hydrogen would fall, after making corrections for Doppler effects caused by motions of the Earth and Sun. Correcting for those motions yields the local standard of rest (LSR in jargon), a frame of reference used by astronomers to mean floating amid the nearby stars. The uncertainty in the LSR is several kilometers per second, which amounts about 10 kHz at this frequency, so the signal could have been exactly at the hydrogen frequency or up to 20 kHz away.

What are the chances that a signal would fall in a channel that says "hydrogen"? Signals could have fallen into any of the fifty receiver channels with equal odds, but only three or so channels would seem to correspond to hydrogen. That's a 3/50 or 6% chance of saying "hydrogen" and a 94% chance it would not. So it's somewhat improbable that interference would fall so near the hydrogen frequency, although not wildly unlikely.

To dismiss the Wow as man-made interference, we would have to believe that a number of pretty unlikely events occurred at the same time: that a secret spacecraft crossed the antenna's beam in just the right way to mimic the transit of a celestial source, and that it was transmitting in the protected hydrogen band, and that its frequency happened to fall almost exactly where the hydrogen frequency would fall in the local standard of rest. Major U.S. reconnaissance satellites such as Big Bird and Key Hole were *not* in the vicinity.

DOPPLER-CORRECTED?

The Wow frequency seems close to the magic hydrogen frequency only after we make some adjustments to compensate for our various motions, and the fact that it comes close to saying "hydrogen" after those adjustments makes the signal's frequency seem especially intriguing.

When we look out into space, everything is in motion and frequencies are shifted higher or lower by the Doppler effect, so we need to tune our receivers higher or lower depending on how fast we are approaching a source or receding from it. The Sun zips along at about 20 kilometers per second compared with the average velocity of local stars—much faster than a rifle bullet—and the Earth shares that motion along with its own motion orbiting the Sun, speeding up as it's pulled near and slowing down as it swings away. The Earth is also spinning, swinging an observer at the equator in a circle at a half-kilometer per second—a thousand miles per hour—although hardly moving at all at the poles. This medley of motion causes Doppler shifts in frequency; the reason we don't usually notice it is that the distance between our transmitters and receivers rarely changes very fast because we're all riding the same merry-go-round.

To get rid of pesky Doppler effects, radio astronomers add up the velocities along the telescope's line of sight, calculate the frequency shift (about 4.7 kHz for every kilometer per second at the hydrogen frequency), and often tune the receiver to make that shifted frequency fall in the center of their dial. Figure 2.3 shows Doppler shifts in all directions at the time the Wow was detected; in the direction the antenna was pointed, hydrogen at rest would be

```
Declination (1950)
 90.0  -81 -81 -81 -81 -81 -81 -81 -81 -81 -81 -81 -81 -81 -81 -81 -81 -81 -81 -81 -81 -81 -81 -81 -81
 80.0  -94 -92 -89 -86 -82 -78 -74 -70 -67 -65 -64 -64 -65 -67 -70 -73 -77 -81 -85 -89 -92 -94 -95 -95
 70.0 -104-100 -95 -88 -80 -72 -65 -58 -52 -48 -46 -45 -47 -51 -57 -64 -71 -79 -87 -94-100-104-106-106
 60.0 -111-106 -98 -88 -76 -65 -53 -43 -35 -29 -26 -26 -28 -34 -42 -52 -63 -75 -86 -97-105-111-114-114
 50.0 -115-108 -97 -85 -70 -55 -41 -28 -17  -9  -5  -5  -8 -16 -26 -39 -54 -68 -83 -96-107-115-119-119
 40.0 -115-107 -94 -79 -62 -44 -27 -11   2  11  16  16  12   3  -9 -25 -42 -60 -77 -93-106-115-120-120
 30.0 -112-102 -88 -71 -52 -31 -12   6  20  31  36  36  32  22   8 -10 -29 -49 -69 -87-101-111-117-117
 20.0 -106 -95 -80 -61 -40 -18   3  23  38  49  55  56  50  40  24   6 -15 -37 -59 -78 -93-105-111-111
 10.0  -96 -85 -69 -49 -27  -4  18  39  55  67  73  73  68  57  41  21  -1 -24 -47 -67 -83 -95-101-101
  0.0  -83 -72 -55 -35 -13  10  33  53  70  82  88  88  83  71  55  35  13 -11 -33 -53 -70 -82 -88 -89
-10.0  -68 -57 -41 -21   1  24  46  67  83  95 101 101  96  85  69  49  27   4 -18 -39 -55 -67 -73 -73
-20.0  -50 -40 -24  -6  15  37  59  78  93 105 111 111 106  95  80  61  40  18  -3 -23 -38 -49 -55 -56
-30.0  -32 -22  -8  10  29  49  69  87 101 111 117 117 112 102  88  71  52  31  12  -6 -20 -31 -36 -36
-40.0  -12  -3   9  25  42  60  77  93 106 115 120 120 115 107  94  79  62  44  27  11  -2 -11 -16 -16
-50.0    8  16  26  39  53  68  83  96 107 115 119 119 115 108  97  85  70  55  41  28  17   9   5   5
-60.0   28  34  42  52  63  75  86  97 105 111 114 114 111 106  98  88  76  65  53  43  35  29  26  26
-70.0   47  51  57  64  71  79  87  94 100 104 106 106 104 100  95  88  80  72  65  58  52  48  46  45
-80.0   65  67  70  73  77  81  85  89  92  94  95  95  94  92  89  86  82  78  74  70  67  65  64  64
-90.0   81  81  81  81  81  81  81  81  81  81  81  81  81  81  81  81  81  81  81  81  81  81  81  81

RA      24  23  22  21  20  19  18  17  16  15  14  13  12  11  10   9   8   7   6   5   4   3   2   1
```

Figure 2.3. All-sky map of Doppler shifts for a source at the hydrogen frequency viewed from the Ohio State site at the time of the Wow detection, with shifts in kilohertz. The Wow position is between the values for right ascension (RA) 19 and 20 hours, near declination -30°. Shifts at declinations below the local horizon are included. Calculations are based on a program by John Ball.

shifted down about 40 kHz compared to its frequency in the lab because we were moving away at the time. The Wow frequency was about 10 kHz lower than that—within one channel of where a signal sent out at the hydrogen frequency would be expected to fall (it did not fall in the center of Ohio's channels because they were also correcting for our motion around the galaxy).

Years later, in observations discussed in a subsequent chapter, hydrogen in the Wow direction was found to be shifted 15-20 kHz lower than where our local motion would put it, presumably due to its own local motion—putting the Wow frequency very nearly where it would fall if its source was amid that distant hydrogen. The fact that the signal was close to the magic hydrogen frequency after these arcane adjustments adds to its interest as a candidate interstellar signal but hardly proves it was one.

In 1998 Jerry Ehman, who first noticed the signal, reported that he believed its frequency was 100 kHz or ten channels higher than originally given, due to a design oddity in the receiver. If that is correct, then the signal is not so remarkably near the hydrogen frequency, although it is still in the hydrogen band.

PECULIAR PATTERN

Another intriguing aspect of the Wow is that several other signals seem to accompany it, forming the peculiar-looking pattern shown in Figure 2.4. Features in channels 4 (maybe), 7, and 16 look like a second signal stepping in frequency, and having the same source as the signal in channel 2 because the intensity of the features increases along with that of channel 2 as the antenna sweeps across the source. The features in channels 7 and 16 are almost certainly real because fluctuations in the background noise would not produce even one peak that strong over several days, but the channel 4 feature is iffy— a four-sigma noise peak is expected in some channel every minute or so, although its intensity does follow the antenna pattern.

Taken together, the features look rather like a pattern well-known to physicists and astronomers: the Lyman series of spectral lines emitted by hot hydrogen at ultraviolet wavelengths, illustrated in Figure 2.5. Some writers have suggested that broadcasters might use patterns such as the series

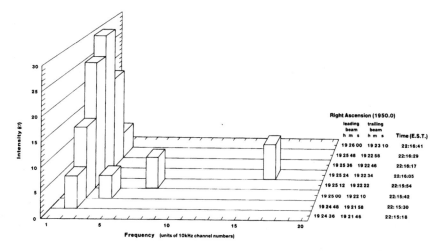

Figure 2.4. Features stronger than 3 sigma in channels 1 through 20. The intensity in channels 4, 7, and 16 was 4, 6, and 7 sigma, respectively, increasing along with the signal in channel 2 as though they had the same source (having a correlation coefficient of r=0.99). Data courtesy of North American AstroPhysical Observatory.

of prime numbers 2, 3, 5, 7, 11, 13, 17... to prove a signal is artificial, and the Lyman series might serve the same purpose because many astronomers would recognize it—and realize that it's out of place in the radio spectrum.

The Wow pattern is not a good statistical match to the pattern of Lyman wavelengths, although the pattern does match the pattern of strengths of the lines if some assumptions are made (a feature in channel 3 undetected just before the beam began sweeping over the source, and a feature outside the band where channel 93 would have fallen). That does not prove anything, because with only a few data points and some convenient assumptions it's not

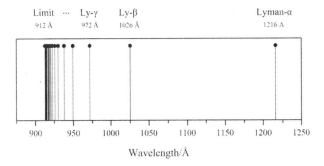

Figure 2.5. The Lyman series of spectral lines in the ultraviolet spectrum. Source: Wikipedia.

too tough to match something, but the peculiar pattern is another facet of the Wow signal that makes it interesting.

WHY SO ELUSIVE?

If the Wow did come from a celestial source, how could Ohio's roughly one hundred additional observations have failed to detect it again? The fact that it was detected in only one of the two beams suggests one answer—that it might have been intermittent.

Most searches for ET have assumed that signals are present all of the time, but that requires broadcasters to either radiate isotropically in all directions continuously, which would be awfully expensive, or to radiate signals toward us all the time with a big antenna, which would be awfully lucky for us.

Transmitting in all directions continuously would require the equivalent of thousands of power plants to reach out 500 light years, so broadcasters might put the signal "on the air" only part of the time. Broadcasting for only one second in ten thousand, for example, would reduce the cost by the same factor saving lots of power plants, but would result in an intermittent signal that would require searchers to listen for several hours to catch.

Transmitting with a highly directional antenna illuminating just one or a few stars at a time is another way that broadcasters might save power, and could result in intermittent signals. The savings are an accountant's delight: only one percent of the output of one big power plant would be needed to signal across 500 light years if focused by an antenna the size of the 1,000-foot radio telescope we have at Arecibo, Puerto Rico, assuming in both examples a receiver using a similar antenna. But a searcher can't see the signal when the antenna is not pointed in his (or hers or its) direction, so the broadcast would appear to be intermittent and would seem elusive.

Intermittent signals might result from something as mundane as rotating planets—for example, fixed antennas sweeping a transmitting beam across the sky once each day. Distant observers would see the signal briefly as the beam swept across them but would have to wait until the source made a complete rotation to see it again. That sort of broadcast strategy might please alien engineers as well as accountants because the antenna would not need to move to track targets. Such signals would be hard to detect with the quick looks of our typical searches, but would repeat periodically.

The Wow might be elusive, then, because it illuminates us rarely for any one of these reasons; Ohio State's follow-up observations might not have looked long enough or at the right time to catch it on a subsequent occasion. And, since their "looks" were daily transits on a periodic schedule, a source that was also periodic might need to come back into synchronization again before it could be detected—and that might not happen for hundreds of days as we shall see later.

An alternative explanation for appearing in just one beam could be a signal drifting in frequency at just the right rate to stay in channel 2 while it was in one beam, but drift through channel 1 undetected and out of the band before the second beam swept past (or drift into the band after the first beam swept past). A drift rate in the range of 100–140 Hz/sec would do the trick. Some drift in the frequency of interstellar broadcasts is expected due to our motion, although only around 1 Hz/sec, and an unknown amount of drift might be caused by motion of the transmitter. A broadcaster might intentionally sweep the frequency across the radio spectrum, but that would make a signal harder to find repeatedly, at least with our present-day searches.

NO HOAX

The topic of aliens is fertile ground for fantastic claims and outright lies. Pictures of flying saucers are common, and many are admitted fakes; some UFO pictures *look* like hubcaps or pie tins because they *are*.

Was the Wow a hoax? I've found no reason to think so.

The reputations of the Ohio State scientists argue against hocus-pocus by observatory staff. John Kraus, founder and director of the observatory, was a prominent figure in the development of radio astronomy. He literally wrote the book on the subject—his *Radio Astronomy* textbook is a standard on many an astronomer's bookshelf, and he also wrote the textbooks *Antennas* and *Electromagnetics*. Bob Dixon, the SETI project director, never sounded a false alarm during decades of supervising the world's longest-running search. Jerry Ehman, who first noticed the signal, had worked on the original Ohio survey and subsequently became a professor at Franklin University. All have been cautious in their claims.

Kraus wrote that "it could have been a signal from an Earth-launched space probe of which we are unaware" and "it could have been a signal from an extraterrestrial civilization... but... this is pure speculation." Ehman regards the origin of the signal as an open question: "... since all of the possibilities of a terrestrial origin have been either ruled out or seem improbable, and since the possibility of an extraterrestrial origin has not been able to be ruled out, I must conclude that an ETI (ExtraTerrestrial Intelligence) might have sent the signal that we received as the 'Wow' source."

The fact that the signal appeared in only one beam and not twice as expected of celestial sources argues against monkey business by insiders. Anyone intending to fake the response of the antenna to a celestial source would know that a signal should appear twice and would presumably fabricate data for both beams to create a more convincing deception.

A prank by an outsider also seems unlikely. It would have been difficult to introduce a spurious but convincing radio signal into the system from outside, and it's unlikely that an outsider could have gotten access to the locked receiver room to fiddle with software or printout. Few outsiders would know how the strength of a signal should rise and fall, or other esoteric aspects of the telescope.

SUMMARY

The Wow signal had several characteristics expected of an interstellar radio signal. Most strikingly, it had the same signature as a celestial source passing through a beam of the antenna, yet the only natural radio sources in the vicinity are a thousand times fainter. The signal had the narrow bandwidth typical of radio signals, yet there was no sign that it was from an aircraft, spacecraft, or other man-made source. The signal's frequency was close to the hydrogen frequency often suggested for interstellar communication, and several weaker signals that seem to come from the same source form an intriguing pattern.

The signal was not seen again in roughly four hours of additional observations at Ohio State over a period of one hundred days, perhaps because the source is not always "on" or not always pointed in our direction or drifts in frequency. A lighthouse-like radio source, for example, might illuminate us so rarely that detecting it even once was lucky, and Ohio State's re-observation attempts would not have had a good chance of detecting such a source a second time.

Confirming the Wow signal as an interstellar beacon (or possibly some novel celestial object) would require it to be seen repeatedly. The following chapters recount my attempts to detect it a second time—the only efforts other than Ohio State's.

Chapter 3

Small SETI Radio Telescope

When I first learned about the Wow! signal, I was already thinking of building a radio telescope to search for signals from other worlds. It was a slightly crazy idea but not full lunacy—there was only one full-time search at the time (Ohio's), and it seemed seductively sensible that more searches would be more likely to detect broadcasts from the stars if any existed. New technologies were appearing that could make a small system very sensitive, and a small antenna would have the advantage of seeing more stars at once than big antennas—viewing the forest rather than individual trees, perhaps able to catch strong signals that big antennas were missing.

I decided to build a small radio telescope to hunt for the Wow signal and other signs of ET. The fact that my education was in liberal arts and urban planning wasn't fatal; I had built radios as a teenager, and my graduate and professional work had drawn me into computer programming and analyzing data. I'd even published a paper about searching for interstellar beacons in a scientific journal, so it seemed perfectly logical to do my own search.

I might have been primed to sally forth on such a potentially quixotic venture by a story told by Ashish Sen, my graduate school statistics professor and advisor. As the story went, a researcher in the early days of gasoline engines noticed a measurement in some samples of gas that implied an almost impossibly good batch—an outlier, in statistics jargon—and instead of ignoring it, he doggedly tracked the path the gasoline had been shipped along, looking for an explanation. He found that the stuff had been contaminated in transit, and the contaminant was developed into a valuable additive that made engines run better. The lesson was that hunting down the cause of mysterious measurements sometimes bags big game.

BUILDING AN INTERSTELLAR RADIO

Radio SETI systems are basically just radio telescopes with special detector gizmos that monitor many different frequencies at once. The radio telescope part can be fairly simple: a dish antenna, some low-noise amplifiers, and a receiver quite like a regular radio except usually tuned to a much higher frequency band. To make it an interstellar search system, you run the receiver's output through a spectrum analyzer instead of a loudspeaker, and the analyzer divides it into many narrow slices of frequency and measures the amount of power in each channel simultaneously, allowing even very weak signals to be detected.

I designed a system using Kraus's textbook on radio telescopes, radio engineering handbooks, and parts catalogs, building everything from scratch the first time around. Building microwave receivers without a lab for testing them turned out to be a dumb idea; hardly anything worked. Seeking help from top electrical engineers, I found that some were willing to donate or build state-of-the-art gear for the project, so all I had to do was connect the boxes and write computer programs to run the system. That took a year.

I needed a big dish antenna and found one through a hamfest—a flea market for amateur radio operators and technicians—a twelve-foot dish with a motorized base to steer it. The dish had been used to relay phone calls between microwave towers, and the base was from a military radar built in 1944 (an MP-61 pedestal from an SCR-584 anti-aircraft fire control radio set, in drab military nomenclature). A twelve-foot dish is small by radio astronomy standards but big to haul down the road; it was a spun aluminum bowl like a wok that couldn't be disassembled. With the help of friends, I lashed it on top of a rented van and drove it into Chicago at dusk to avoid police attention, with tractor-trailers steering clear of our oversize load. At the end of the trip, we rolled the big dish on edge through narrow alleys to the considerable astonishment of neighbors.

Radio telescope receivers are more sensitive than regular radios because they use a low-noise amplifier to boost the incoming signal, adding very little noise. Hiss from the first amplifier gets boosted roughly a million times in later amplifiers and will drown out weak signals if too loud; quiet amplifiers are highly prized because reducing noise from the receiver increases a telescope's sensitivity—cutting the noise in half has the same effect as doubling the area of the antenna but is much cheaper. William "Tap" Lum was the leading LNA builder, practicing his esoteric electronic art at the Radio Astronomy Laboratory at University of California at Berkeley. He had recently built twenty and sent the best ones to the Arecibo Observatory in Puerto Rico; he sent me a discard that was better than anything ready money could buy.

The rest of the receiver was built by Skip Crilly, a designer of frequency synthesizers at the Hewlett-Packard company. It was custom-designed and very advanced, using logic circuits to precisely synthesize frequencies, and tunable by computer. The receiver's job was to convert a 1420 MHz incoming signal (a voltage varying 1420 million times per second), down to a frequency in the audio range (varying a few thousand times per second) where it could be analyzed. The trick in radio is to heterodyne or mix an incoming signal with one from a local oscillator, which produces new signals at the sum and difference frequencies—the latter lower frequency being easier to work with. Skip did it in stages: a 1400 MHz local oscillator (a 10 MHz crystal in a temperature-controlled oven, multiplied up in frequency) was first mixed with incoming signals around 1420 MHz to get a 20 MHz intermediate frequency, then a fine-tunable local oscillator signal near 20 MHz was synthesized to drop the frequency down to the audio range of a few kilohertz.

What made this contraption an interstellar radio receiver suitable for hunting ET was a spectrum analyzer with many narrow channels, able to monitor many frequencies simultaneously. Such gizmos are thought to be the key to detecting radio signals from other worlds for several reasons. First, they can listen at many frequencies at once, so searchers don't need to tune across the dial one frequency at a time which would take too long. Second, dividing the radio spectrum into narrow slices reduces the amount of noise in each channel, making them more sensitive to signals that are expected to be narrow. Third, extraterrestrial signals should drift slightly in frequency due to the Earth's rotation, and narrow channels can reveal that drift. The Hewlett-Packard company was selling just the thing for $10,000; their suit-case-sized HP3582A could monitor 256 channels simultaneously, display the spectrum, and send its findings to a computer to be analyzed. Barney Oliver, vice-president of research and development at HP (and later head of NASA's SETI program) and factory manager Charles Kingsford-Smith arranged the donation of a smoke-damaged but working one. It provided better spectral resolution than any radio telescope anywhere at the time; later versions of the system had up to 32,000 channels as computer technology improved.

The ability of spectrum analyzers to find signals hidden in noise is amazing. Figure 3.1 shows a few hours of real sky data, with a test signal added in just one channel. The hiss from the galactic plane passing through the antenna's beam makes a big bump when all channels are averaged together, and the test signal in a single channel is completely undetectable until each channel is examined individually. Then the signal pops up and the noise from the galaxy all but disappears because so little falls in each channel.

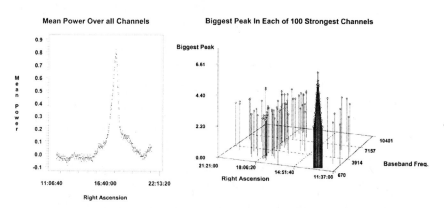

Figure 3.1. Radio noise from the galaxy (left), concealing a simulated signal (right). The chart on the left shows the rise and fall of hydrogen as the Milky Way transited the antenna, averaging all channels in a 8,192-channel version of the receiver. A weak test signal was added to one channel to simulate a drifting narrowband signal but is not noticeable. The chart on the right shows the channels with the most power, and the simulated signal stands out clearly, although it is only about twice as strong as the biggest noise peaks. A slight drift in frequency can also be seen, about 6 Hz/minute.

The system needed to be automated to be practical so I computerized everything with a CompuPro microcomputer featuring a 2-megahertz 8-bit Intel 8085 processor, two eight-inch floppy disk drives that could record one megabyte each, and 64 kilobytes of memory—industrial strength hardware for its day. William Jayne and Terry Lilley built the computer, soldering specialized circuit boards to make it affordable.

Programmed in the Fortran language (sensibly named for "formula translation"), the computer calculated the frequency settings needed to compensate for Doppler effects using three pages of trigonometry, tuned the receiver, and controlled the spectrum analyzer to collect and average spectra. Data was harvested from the analyzer every minute and stored on floppy disk, then later charted to look for signals. The system typically monitored a 10-kHz chunk of spectrum with channels 40 Hz wide but could zoom in to examine signals with 1 Hz or higher resolution to look for the drifting frequency signature of extraterrestrial signals.

To observe the Ohio State position without having to track it, I pointed the antenna about 21 degrees above the horizon and let the Earth's rotation sweep the beam across the sky. The beam swept across the Wow locale once each day, as well as the star-rich plane of the galaxy, monitoring objects for the half-hour it took them to drift through the three-degree-wide beam—a spot on the sky six times as wide as the Moon—and in the course of a day the beam swept across 1% of the sky. I figured that if I let the contraption run long enough, it might catch the Wow or perhaps find something else interesting; the system saw more sky for longer each day than other searches using bigger antennas, although it was not as sensitive.

Looking for Aliens

The system saw its "first light" in 1983, detecting the radio Sun. Running for between six and 24 hours daily, it detected the Milky Way each day as the beam swept across the sky, proving that the system worked well enough to see interstellar hydrogen thousands of light years away.

Each day I made charts like the example in Figure 3.2, looking for vertical bumps or spikes that might indicate a strong signal, and more detailed charts looking for narrowband features. Lots of signals showed up, some intriguing, but none with signs of an interstellar beacon such as the Earth's Doppler drift, and none from the Wow locale.

The system's best sensitivity using 1-Hz channels was good enough to detect a transmission from a 300-foot antenna fueled by a big power plant at a range of up to 50 light years (about 10^{-21} W/m² in 1-Hz channels)—a volume of space containing hundreds of stars. The range would be greater for more powerful transmitters or larger broadcasting antennas but less with the 40-Hz channels I used for most observing runs.

The most intriguing event I ever encountered was a fairly strong signal less than 30 minutes after the antenna swept across the Wow locale, showing the rise and fall intensity signature a celestial source, shown in Figure 3.2. Using programs that calculate the motions of solar system objects, I found that the Moon had been near that position at the time, perhaps reflecting terrestrial interference back into the antenna (it was not in the vicinity when the Wow was detected).

The system ran every night through much of the mid- and late- 1980s, and for months-long periods in the 1990s. By 1998, local interference, failing hardware, and more ambitious ideas about how to search added up to retirement for the small SETI radio telescope.

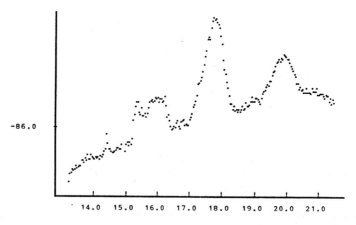

Figure 3.2. A transit of the Milky Way at right ascension 17ʰ 45ᵐ on May 19, 1984, and something else at about 19ʰ 50ᵐ—probably the Moon. The horizontal scale is right ascension in hours and the vertical scale is noise power. The bump between 15 and 16 hours does not display the antenna pattern curve and is probably interference.

Green Bank

In May of 1985, I was invited to give a talk about my project at a workshop sponsored by the National Radio Astronomy Observatory in Green Bank, West Virginia.

Green Bank is a community of fewer than 200 souls in Pocahontas County—a five-hour drive from Washington D.C. on twisty two-lane roads through lush Appalachian mountain valleys. There are few settlements along the way; most people live in modest houses, trailers, or shacks alarmingly near the road, whose main appeal appears to be distance from neighbors. The drive through the woods is lyrical and hypnotic; it's a road that hundreds of radio astronomers have traveled along with local good ol' boys. The remote location was selected because the terrain and sparse population frustrates broadcasting, which often interferes with radio astronomy; state law guarantees freedom from interference for ten miles in every direction.

The observatory is just outside Green Bank, its entrance marked by historic antennas built by radio astronomy pioneers Karl Jansky and Grote Reber. Through the gate and down a private road is a low administration building and electronics laboratory on the left, and a two-story dormitory-like building for visiting astronomers on the right. The road ends in a barrier; to get closer to the big antennas you need to use an observatory diesel car, bicycle, or shoe leather.

Largest of the Green Bank telescopes at that time was the old 300-foot dish, a towering lattice of steel too big to steer. It was simply tipped to the angle above the horizon where a target would transit, and the Earth's rotation swept its beam across objects of interest. The antenna collapsed into a tangled pile of scrap metal in November 1988 due to the failure of a gusset plate, the same flaw that sometimes brings down bridges. A bigger, fully steerable, and much more rugged antenna was completed in 2001 to replace it, called the Green Bank Telescope or 100-Meter (about 330 feet across, in old-fashioned units). On more formal occasions it's called The Robert C. Byrd Green Bank Telescope, in honor of the West Virginia senator who was the chairman of the Appropriations Committee for many years.

The other major telescope was the 140-foot, the biggest polar-mounted antenna in the world. Gleaming white and rising high above the landscape on a winged concrete pedestal, it looked oddly out of place—as though it might have been built to cruise the seas rather than stand on land. The oldest antenna was the 85-foot, which Frank Drake used in 1960 in the first modern search for radio signals from two nearby stars.

The workshop consisted of three days of talks delivered by several dozen scientists active in SETI, reporting their recent work and talking shop at meals and in the evenings. I described the small radio telescope system I'd built and my early results, touted advantages of small systems, and also mentioned some half-baked ideas. Big mistake; sloppy thinking is blood in the water to scientists and engineers. A few tough questions followed, dry-mouth struck, and I croaked lame answers making a poor showing.

One consolation was getting acquainted with Paul Horowitz, who had recently built an 8-million channel spectrum analyzer at Harvard—the first to use a huge number of very narrow channels, and in some respects a nearly perfect interstellar radio receiver. Paul hitched a ride with me out of West Virginia, and in later telephone conversations I tried to interest him in using his magical machine to look for the Wow. Being a busy fellow and clever enough to build interstellar radios, he countered, "Why don't *you*?"

Chapter 4

META

Hypothesis: ultra-narrowband?

A lot of smart folks think that interstellar beacons would send very narrow signals—more like the pure tone of a tuning fork or Morse code than the wide range of frequencies in speech, music, or television. Paul Horowitz had a receiver that listened on eight million ultra-narrow channels in the same band as the Wow! signal, which seemed like the perfect gizmo for finding the Wow if it was very narrow—possibly a weak signal that only occasionally got strong enough for Ohio to detect. The bandwidth of the signal was unknown except that it fit in one Ohio channel; it could have ranged from a pure narrowband tone less that 1 Hz wide, or up to as wide as 10 kilohertz like an AM radio station blaring salsa music.

Paul's receiver also covered a much wider range of frequency than mine (400 kHz vs. my 25 or less), and my small system might have been missing the Wow signal by listening at the wrong frequency. Best of all, he had a big antenna that could track sources for hours, perhaps long enough to catch an intermittent signal.

THE SECOND FULL-TIME SEARCH

Paul Horowitz got his amateur radio operator's license at age eight and was now teaching Harvard graduate students how to design and build electronic instruments for exotic research projects. His book *The Art of Electronics* co-authored with Winfield Hill is widely used, engaging for the genre, and occasionally even funny.

Horowitz got into the SETI business in 1978, when he observed several hundred stars at Arecibo in one of the most sensitive searches ever done. The huge collecting area of the 1,000-foot antenna made the search very sensitive, but a special trick made it extraordinarily sensitive. The trick was to listen on such narrow slices of frequency that the narrowest possible signal from a beacon would fall in a matching narrow receiver channel, so the signal's power would not be diluted by being spread over the millions-times-wider channels used in most radio astronomy receivers. Existing receivers could not do that, so he recorded the receiver output on magnetic tape and later played the recording through a software transform that divided it into 65,536 channels each just 0.015 Hz wide. In the early 1980s, Horowitz built a portable version of his spectrum analyzer dubbed "Suitcase SETI" that could do it in real time and conducted additional searches at Arecibo.

The advantage of his strategy was high sensitivity to signals with razor-thin bandwidth like pure tones; the drawback was that his spectrum analyzer could cover only a narrow band of frequency—only 1 kHz, one-tenth as wide as an AM radio station. If ET failed to tune his/hers/its transmitter to make a signal fall in that thin band, it would be missed. What was needed was a receiver covering a wider band with millions of channels, but no such thing existed. He decided to build one and managed to get the 84-foot Harvard-Smithsonian radio telescope outside of Boston dedicated to a full-time search, pictured in Figure 4.1.

META

In 1985, Horowitz built an 8-million channel spectrum analyzer named META, for Mega-channel ExtraTerrestrial Assay. It was funded by the

Planetary Society with a $100,000 gift from Steven Spielberg (director of the science-fiction film *E.T. the Extra-Terrestrial*), and covered a 400 kHz band with 0.05 Hz resolution—wide enough to catch signals with a variety of Doppler shifts due to motion at the signal's source, and signals not aimed precisely at our Sun.

The META spectrum analyzer lived in two equipment racks about the size of a big refrigerator, stuffed with 144 book-size circuit boards, each having its own Motorola 68000 microprocessor, which was the same chip as used in early Apple Macintosh computers. Each processor calculated the spectrum for a chunk of the band, and hundreds of colored lights flashed across the front panel showing progress and impressing visiting journalists.

One problem with using a comb of very narrow channels is that a slow Doppler drift caused by the Earth's rotation sweeps extraterrestrial signals across them so quickly that not much energy gets accumulated in any one

Figure 4.1. The 84-foot Harvard-Smithsonian Agassiz Station radio telescope with six-year-old Jake Horowitz. Courtesy of Paul Horowitz.

channel. That would be fatal except for a clever trick: the receiver's frequency was slowly slewed in frequency or "chirped" at the same rate as the Doppler drift, so a drifting extraterrestrial signal would stay in a similarly drifting channel, while local signals at a constant frequency were smeared across many channels, making the system insensitive to interference.

META's antenna could track objects if an operator was present, but none usually was; it was more convenient to let the Earth's rotation sweep the half-degree-wide beam across the sky every day. The antenna's elevation was changed by a half-degree daily to cover a new ring of sky, and after several hundred days the two-thirds of the sky visible from the site would be surveyed (a META II system in Argentina later surveyed the southern hemisphere sky). Celestial objects passed through the beam in about two minutes, and strong signals were logged by the computer for later review; every so often, Horowitz would track such promising positions for a while to see if the signals were still there.

OAK RIDGE

I arrived in Boston in August of 1987, and Paul and I drove out into the countryside through historic New England towns like Lexington and Concord to the Harvard-Smithsonian Astrophysical Observatory 30 miles west of Boston. The observatory site is on the wooded hilltop of Oak Ridge, near the small town of Harvard founded in the early 1700s and home to some 5,000 souls (the university is located back in the Boston area).

Harvard's observatory consisted of an old Ivy-league limestone and leaded-glass building, a separate domed structure housing the 61-inch Wyeth telescope—"largest optical telescope east of Texas"—and several smaller telescopes. Paul notes that Frank Drake, who conducted the first SETI observations in 1960, worked there while a graduate student in 1955, refurbishing the then-old optical telescope and greasing gears on the new radio one when it was built in 1956. The 84-foot (26-meter) radio telescope was located down the hill through the woods.

The dish was set on a tall tilted steel pedestal topped with an equatorial mount pointed at the North Star. Once the dish was raised to the desired elevation, electric motors slowly swung it in an arc across the sky to counteract the Earth's rotation and track celestial objects. A long rectangular tail jutted toward the ground from the back of the dish, containing a steep stairway up into the bowl for access to receiver equipment. A single-story cinder block building beneath the big dish housed a control room, a small electronics lab, and a tiny room with a cot for visiting observers.

Paul showed me how to point and steer the dish and how to run the META computer, and soon we were tracking one of the two Wow positions and recording occasional events that might prove to be interesting signals. He left me to run the system by myself for the next four days—the best interstellar radio receiver on the planet, pointed in the direction of the best candidate signal ever seen.

I had a very simple plan: track one of the two Wow positions on first day, track the other position on the second day, then repeat. If a signal appeared at one spot both times and not at the other, it would be a sign of celestial origin and tell us where to look harder—not proof of little green guys, but evidence that something might be out there and probably not just local interference. I tracked each position for as long as possible to increase the chances of catching intermittent signals.

To track things with old-fashioned pre-computer-controlled antennas like the one at Oak Ridge, you first turn a big black knob to energize an electric motor that slowly raises the dish, then turn it off when a dial shows that it's elevated to the declination of your target—in this case, about 20 degrees above the horizon. You can't see much of the big dish from the control room beneath it, so you have to trust the dials. If the target happens to be in front of the dish at that moment, you throw a tracking switch to make the dish slowly arc across the sky at the almost imperceptible rate of one revolution per 23 hours and 56 minutes to follow the object. But usually you need to slew the dish along an east-west arc until you get to the so-called hour angle where your target is currently located (the amount of time the object is away from the meridian overhead) and then throw the switch to start tracking.

Since the Wow positions are above the horizon at Oak Ridge for only about four hours a day, I had to lower the dish until its edge was only a few feet above the ground so I could begin tracking as it rose above the horizon, then follow its arc up and back down until it set four hours later—nearly

baseline

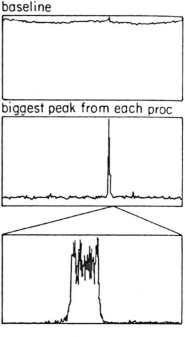

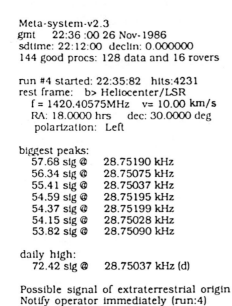

Meta-system-v2.3
gmt   22:36 :00 26 Nov-1986
sdtime: 22:12:00 declin: 0.000000
144 good procs: 128 data and 16 rovers

run #4 started: 22:35:82   hits:4231
rest frame: b> Heliocenter/LSR
    f = 1420.40575MHz   v= 10.00 km/s
    RA: 18.0000 hrs   dec: 30.0000 deg
    polarization: Left

biggest peaks:
    57.68 sig @   28.75190 kHz
    56.34 sig @   28.75075 kHz
    55.41 sig @   28.75037 kHz
    54.59 sig @   28.75195 kHz
    54.37 sig @   28.75199 kHz
    54.15 sig @   28.75028 kHz
    53.82 sig @   28.75090 kHz

daily high:
    72.42 sig @   28.75037 kHz (d)

Possible signal of extraterrestrial origin
Notify operator immediately (run:4)
Possible signal of extraterrestrial origin

biggest peak from each proc

Figure 4.2. Example of META's console display. The top graph shows the full 8-million channel spectrum with each pixel on the screen representing the average of thousands of adjacent channels (displaying all 8 million channels with one pixel for each channel would require a screen more than a mile across). The small bump in the center of the spectrum is the hydrogen line. The middle graph shows the biggest peak in each of the 128 processors crunching a chunk of the band. The bottom graph zooms in on a possible signal showing individual channels to see if the signal is really narrow (one or two channels) or spread over many channels, like the interference shown here. Courtesy of Paul Horowitz.

driving the dish into the ground in order to track for as long as possible. I did that for four nights, racing outside in the dark with a flashlight to make sure the dish didn't crash.

I stared at the console during each observing run, watching graphs like the examples in Figure 4.2 update every 23 seconds. The system recorded some information for the strongest signals, but very little to avoid filling up the tiny disk drive. After each night's observations, I did a quick analysis on graph paper to see if I'd caught anything obvious—strikingly strong signals, or weaker ones repeating at regular intervals, or anything not random. After four days, I returned the system to Paul's systematic search, left a bottle of Champagne, and raced for the airport.

ANALYZING SETI DATA

For the next few months I looked at the data every which way I could think of; experiments that take only a few hours to do can take months to digest. META made over a billion measurements each hour, but recorded only a tiny fraction of the data because its disk drive was ridiculously small by modern standards—about 10 megabytes, not enough to store this book—and only a small part of that was available for recording data. I had to set the threshold for recording a "hit" at nearly twenty times stronger than the background noise, which is very high, but noise alone produced a few such peaks every hour. Insisting that signals be so strong is asking a lot (an AM radio signal just ten times the background hiss is strong enough to entertain you if you aren't fussy) but the huge number of channels forces high thresholds to avoid too many false alarms. If we knew the right frequency and could listen on just that one channel, much weaker signals would do.

I found some big peaks in the spectrum—some over 20 times the noise level, shown in Figure 4.3, but none astoundingly big like 50, 100, or 1,000 times. How to decide if any peaks were signals instead of noise? Statistics is the usual gambit. The noise in gizmos like META that don't average their

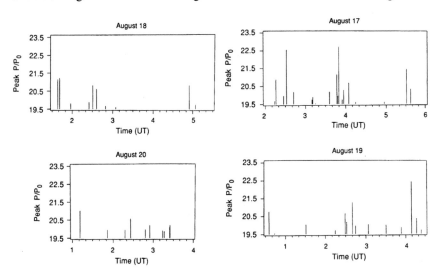

Figure 4.3. The biggest peaks recorded at the Wow locale with the META system. The vertical axis is power, and the horizontal axis is the time each peak was recorded. Graphs on the left are for one position, on the right for the other. Source: Gray, 1994 in *Icarus*, used with permission of Elsevier.

measurements over time follows an exponential distribution, and the biggest peak expected from noise in any number of samples n is easy to calculate as $\ln(n)$ times the average power in a channel ("ln" means natural or base-e logarithm, as opposed to "log", which means the common or base-10 logarithm, sometimes pronounced the same; go figure). Since the system generated 8.4 million samples every 23 seconds, 5.9 billion were generated during a four-hour observing run, and the natural log of that is 22.5—most of my runs had peaks no higher. The biggest peak in all of the runs was 23.3 times the noise, and a peak *that* large was expected at some point during 19 hours.

It's amazing how precise statistics can be! If your data don't differ much from what's expected of noise, then it's hard to get excited.

But real signals *could* lurk among or below big noise peaks, perhaps with regularities noise would not display, so I also looked for peaks repeating at the same frequency, or repeating at regular time intervals, or anything else that random numbers should not do. No patterns could be found to suggest a signal, so the Champagne had to wait for another occasion.

Why did META fail to find the Wow signal? It's possible, of course, that no signal was present. But META was designed to detect only one special flavor of signal—ultra-narrowband and Doppler-corrected—so failing to find that merely proved *that* type of signal was not there while I was looking. Calculations showed that a signal wider than about ten Hz would not have been detected by META, yet the Wow could have been a thousand times wider and still fit into just one Ohio channel. The guys broadcasting might not agree with our guys that ultra-narrow signals are smart.

A fairer test, I realized, would be to use a receiver more like Ohio's to eliminate channel size as a possible problem. Several other big radio telescopes could by this time mimic Ohio's receiver by using spectrometers with many channels of adjustable bandwidth.

META'S FINDINGS

The META search ran from 1985 through 1994 and detected several dozen apparent signals that looked promising, although none were seen repeatedly. Paul Horowitz and Carl Sagan published a report on the first five years of results, saying that they could find no natural explanation for some of the signals and noting that many were clustered near the galactic plane where stars also cluster, just as might be expected if the galaxy was teeming with interstellar beacons or some unknown kind of natural emission. In 2002, T. J. Lazio, Jill Tarter, and Peter Backus re-observed nine of the 11 best META candidate positions using the more sensitive 140-foot telescope at Green Bank and found no sign of continuously-shining beacons, although they only listened for five or ten minutes at most positions.

META only had enough time to detect a signal once during the two minutes its beam swept across sources, and that risked being misled by a single big measurement because digital circuits with lots of memory and processor thingies can get "flipped bits" from passing cosmic rays and other rare events—sometimes turning little numbers into big ones. META's successor, BETA, was designed to be harder to fool in this respect and to cover much more frequency; it's described in a later chapter.

Chapter 5

Very Large Array

Hypothesis: dim but constant?

I wondered if a dim but constant signal might lurk at the Wow! locale—"on" all the time but too weak for Ohio State or META to detect. In that scenario, the signal Ohio detected might have been an occasional high-power blast shouting HEY THERE! to attract attention to a weaker but constant broadcast, or possibly a weak source briefly twinkling brighter.

I discussed this with Patrick Palmer, an astronomer at the University of Chicago who had participated in a search of over 600 stars using the NRAO 140-foot radio telescope, and he suggested I try the Very Large Array in New Mexico—the most powerful radio telescope on Earth. I read up on the sophisticated gizmo and found that it could detect very weak signals by tracking with an array of big antennas and integrating or averaging the receiver output over time, like making a long-exposure photograph. With some coaching from Pat, I wrote a proposal requesting four hours of precious observing time, and to my surprise the gates of Big Science opened and I was invited to test the idea.

Figure 5.1. A portion of the VLA in its most compact configuration. Nine antennas in one arm recede into the distance, while parts of the two other arms extend to the left and right; most of the nine antennas in each of those arms are outside the frame. Each antenna is roughly the size of a 10-story building and weighs 200 tons. The telescope has just finished searching for the Wow signal. Photo: Robert Gray and Sharon Hoogstraten.

THE VLA

The VLA is an array of 27 steerable dish antennas, each 25 meters (82 feet) in diameter, all linked electronically to give the spatial resolution of a single huge antenna up to 36 kilometers across. Completed in 1980 at a cost of $78 million, it lets radio astronomers see the sky in detail as good or better than big optical telescopes. Part of the array is shown in Figure 5.1.

Unlike most earlier radio telescopes, the VLA makes images of the radio sky. Previously, radio sources were visualized using wiggly lines drawn by strip-chart recorders or contour maps, and resolution was usually so poor that a radio source's position might cover many stars and distant galaxies. The VLA can make images good enough to precisely match radio with optical objects, as in the example shown in Figure 5.2.

Nine antennas are parked at various stations along three railroad tracks arranged in the shape of a Y; the rails run off into the distance from a central building, two for 21 kilometers (13 miles), the third for 19 km. In the largest configuration, austerely named "A", antennas are stationed along the full length of each arm and provide a spatial resolution of 0.1 second of arc (one second of arc is 1/3600 of a degree, arcsec in the jargon)—better than the 1-arcsec spots that stars make in conventional ground-based optical telescopes. In the most compact configuration, the antennas are all clustered within less than one kilometer, giving a lower resolution that's better for seeing extended structures like clouds of interstellar gas.

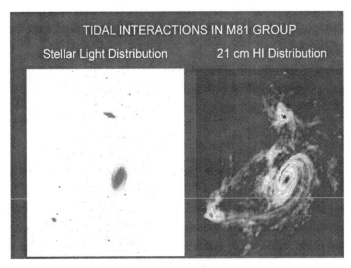

Figure 5.2. Optical and radio images of the same area of sky. An optical image of the M81 group of galaxies is shown on the left as a negative image, and on the right is a VLA radio image of the same field showing hydrogen gas with streamers connecting the three interacting galaxies. Image courtesy of NRAO/AUI. Investigators: M. S. Yun, Paul T. P. Ho, and K. Y. Lo.

Making images of the radio sky relies on two special tricks: aperture synthesis, which simulates an antenna as wide as the entire array, and image reconstruction, which transforms the raw data into images.

Sir Martin Ryle invented aperture synthesis, earning a Nobel Prize in 1974 along with Tony Hewish. The technique exploits the fact that as the Earth turns, an array of antennas moves through many different positions with respect to a celestial source, sampling the radio waves falling on the Earth at many different points. The antennas are interconnected in pairs to form many interferometers, and as the Earth turns, the distance traveled by incoming waves changes between each pair of antennas. That makes the combined amplitude of the signals alternately increase and decrease—wave crests arriving at the same time reinforce each other and become stronger, while crests and troughs interfere and cancel. These interference fringes are what the system actually "sees" and records over many minutes or hours to simulate what a single huge antenna would have seen.

Image reconstruction turns the interference pattern of radio sources into an image using lots of calculations featuring a two-dimensional fast Fourier transform (as opposed, of course, to the slow one), which is somewhat like transforming patterns of concentric ripples in a pond back into the places where pebbles fell.

Most important for my purposes, the system could divide a band of frequency into channels with a variety of custom widths—such as 127 channels each 6.1 kHz wide, roughly comparable to Ohio State's channels.

SOCORRO

The town of Socorro, New Mexico, is sixty miles south of Albuquerque through scrub desert and low mountains. It's a fairly small town, with a population of 9,000, and is very old—Spaniards established a mission there in the early 1600s, and in the 1800s it was a regional mining center. The Rio Grande cuts through, more sand than water during the dry spell when I visited in September 1995. Socorro hosts the New Mexico Institute of Mining and Technology, and on the edge of campus is the Array Operations Center.

The AOC is a modern two-story red and sandy white stucco office building surrounded by unexpectedly green grass and trees. Inside, staff and visiting astronomers use powerful computer workstations and specialized software to turn raw data into knowledge, and electronics wizards in the basement build liquid-helium-cooled receivers and other exotic stuff. Run by the National Radio Astronomy Observatory, astronomers anywhere can apply for a grant of observing time, and it's all free.

I arrived two days early to learn the ropes, nearly a year after applying for time. The observatory supports hundreds of visiting pilgrims from around the globe each year, some new to the telescope like I was, and it's a marvel of efficiency. The reception desk directs you to a temporary mailbox already labeled with your name where you find keys to the office building, keys to a motel-like guesthouse, the password to your computer workstation, and a disturbingly detailed brochure on surviving rattlesnake bites. You are soon introduced to your official "staff friend" who shows you some details of programming the telescope and reducing raw data into images.

Kevin Marvel, then a doctoral candidate—"Captain Marvel" to pals for mysterious reasons—was my staff friend and introduced me to *AIPS*, the Astronomical Image Processing System software package needed to work with VLA data, and to the idiosyncrasies of my Sun UNIX workstation. I spent the days before my observing run studying *The AIPS Cookbook*, a collection of computer instruction recipes for common tasks, trying them out on data from Kevin's research.

I got odd glances from some staffers as I toiled at my workstation, probably because no amateur had ever observed at the VLA before, and nobody had ever used the full array to look for little green men. Location scouts for the science fiction movie *Contact* had recently toured the place, apparently leaving some folks unsure whether my business there was fact or fiction. I wasn't entirely sure myself.

PLAINS OF SAN AUGUSTIN

The actual antennas are 50 miles west of Socorro, surrounded by low mountains in the flat desert Plains of San Augustin. Observers don't really need to travel to the site because a staff operator runs the telescope using stored instructions and sends data tapes back to the operations center daily, but I wanted to see how the system worked and to watch for any source of interference that might contaminate my data—perhaps a Goodyear blimp passing by or some other improbability.

Kevin and I drove out to the telescope site in late afternoon along a hilly two-lane road, through the tiny town of Magdalena, finally descending onto the Plains. The antennas stretched out toward mountains in three directions—just white dots on the landscape below. In the array configuration at the time (called BnA, a hybrid of the two largest spacings) the antennas were scattered across the landscape, and it was impossible to see them all in a single glance. After driving for a while longer, we finally got close enough to an antenna to appreciate its size: taller than a ten-story building, shown in Figure 5.3. We drove past several more on the way to the observatory building, but most of them remained specks in the distance.

We arrived at the two-story concrete block observatory building near sundown and chatted with the operator in the control room until the observatory computer terminated the prior observer's run at the scheduled moment and my program took control. The few antennas close enough to see in the dusk

Figure 5.3. A single VLA antenna, 25 meters in diameter. The railroad tracks in the foreground are used for moving antennas from one observing station to another every few months, changing the array configuration for different kinds of observations. Photo: Robert Gray and Sharon Hoogstraten.

began moving almost imperceptibly toward my first target, and for the next four hours I watched the screen of a spectrum analyzer displaying raw signals piped back from the antennas. No images were made during observations; the hardware ran flat-out just recording the torrent of data flowing in. The telescope spent 20 minutes tracking each of the two fields needed to cover the Wow coordinates, and shorter times looking north and south to cover uncertainties in the positions. Almost half of the precious observing time was spent pointing at known radio sources to later calibrate or adjust the telescope's performance.

Exactly four hours later, my program stopped and the antennas moved on to some other project. We drove back to Socorro at midnight with the data recorded on tape, and for the next three days I worked around and against the clock looking for signals, sleeping a few hours when I got too tired or confused to go on. The specialized computers and software needed for the work were not available in many other places, so if there was something interesting in the data I had to find it quick.

CONTINUUM SOURCES

Making pictures out of radio waves takes a lot of work. The data reduction process includes calibrating the raw data to compensate for imperfections in the instruments, editing it to discard anything screwed up by interference or equipment problems, and finally converting data into images with fancy math. AIPS software does most of the work.

The first images you make are called channel Ø images in VLA jargon, averaging the radio flux from the middle 75% of channels to reveal continuum sources radiating across the entire band. Figure 5.4 shows the sources found in the two adjacent fields containing the Wow coordinates. The telescope's so-called primary beam saw a field of sky about a half-degree in diameter, but the synthesized beams calculated by software were much smaller—depending on the distance between antennas rather than their size—and showed several radio sources within each primary beam.

Some weak sources fell near the two Wow positions, but the VLA is so sensitive that it can detect a million weak sources across the sky, so a few weak natural sources were expected in every field. Three or so were detected in most fields, and since the uncertainty or error box in the Wow positions

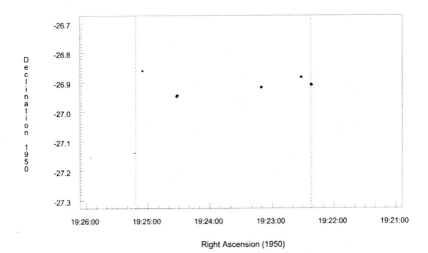

Figure 5.4. Continuum features and speckles of noise in 127 12.2-kHz channels all superimposed, for the two adjacent fields containing the Wow positions, together about two Moons across. The Wow right ascensions 19^h 22^m 22^s and 19^h 25^m 12^s are shown by dashed vertical lines.

covered about 1/12 of each field, the probability of a source falling near those positions by chance was about 3 x 1/12 = 0.25 or one chance in four.

Continuum images can't reveal whether sources are natural or artificial, and would probably fail to show a signal present in just one channel because its flux would be averaged with all the other channels and be reduced to a small fraction of its true value. I needed images for each of the channels to see if a signal was lurking in just one or a few of them.

DATA CUBES

To look for radio signals in single channels, you make a separate image for each channel, which can take hours of number crunching. The result is called a data cube—a computer file of numbers that can be visualized as a stack of images on glass plates that you can look through, with each layer showing the sky at a slightly different frequency. Figure 5.5 shows a data cube for one field, with the continuum sources appearing as lines through the cube because they appear in every channel at the same coordinates on the sky,

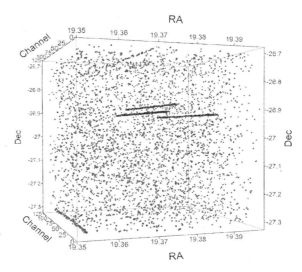

Figure 5.5. A data cube showing the brightest pixels in the 19^h 22^m field. The horizontal axis is right ascension (shown increasing left to right, although the convention is to show it decreasing), the vertical axis is declination, and frequency extends into the cube.

while noise and single-channel features appear as points scattered in both position and frequency.

You can get the spectrum of a source by "drilling" through the cube at its sky position and charting the strength of the flux along that line. Figure 5.6 shows the spectrum for one of the sources near a Wow coordinate; both this and another source near Wow positions showed emission over all channels like natural sources. A big spike in a single channel would be a sign of a radio signal, and might be time to chill the Champagne.

I made data cubes for each field of sky I'd observed, generating over a billion numbers; the challenge then was to discover if there was one sweet spot somewhere in that billion that might be the Wow signal. A single big number would not be good enough; a real source would make a pattern of numbers with the Gaussian bump-under-a-sheet signature of the synthesized beam on the sky. I viewed the cubes on a computer monitor image-by-image and channel-by-channel, looking for bright color-coded patterns amid millions of multihued noise speckles.

For a few minutes, once, I thought I had it. After staring at images all night, at dawn I found myself gazing at an incredibly bright spot at the center

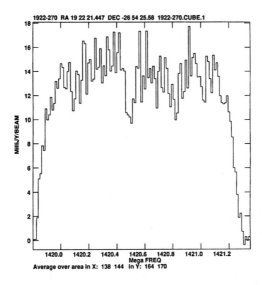

Figure 5.6. Spectrum of a source near one of the Wow positions. Vertical bars show the power (in thousandths of a jansky) at each frequency along the horizontal axis. Source: Gray and Marvel, *Astrophysical Journal* 2001, reproduced by permission of the AAS.

of the screen, radiating six prominent spikes in the pattern of an amazingly strong source. Stunned and tingling, I went outside in the early morning light and sat on a bench to think about it. After a few minutes, I returned to the computer to ask some questions of this shining source: How bright are you? What's your frequency? Where exactly in the sky? I soon found that I'd accidentally displayed the wrong image file—a model of how the telescope beam was *supposed* to look when pointed at a strong source, created by the software during calculations and not a real source at all.

I worked in Socorro until the last minute and then raced for the airport in Albuquerque to fly home with a heavy load of printouts. The Wow had not surfaced in this quick expedition, but astronomical data need lots of sifting to strain signals out of noise. I needed to look harder for weak signals in the billion numbers I'd harvested.

NARROWBAND FEATURES

Over the next few months, I dug deeper into the data using Pat Palmer's Sun Sparcstation named "oskar" on weekends in his office at the Enrico Fermi Institute and using my own computer to analyze some of the results.

With a ton of data (not an official metric unit; about 10,000 megabytes), I needed to automate the process of looking for narrowband features, and did it with an AIPS software tool with the unfortunate name SAD, for "search and destroy."

SAD is a program that automatically identifies radio sources. Given an image file, it searches for areas of flux that are brighter than the background and that have the Gaussian bump-under-a-sheet shape of the beam. Then it produces a list of those apparent sources with their coordinates and flux. It's like a computer solution to the Where's Waldo puzzle—a digital image file goes in, and a catalog of astronomical objects comes out (graduate students used to get degrees for doing that sort of thing). The "destroy" part refers to automatically subtracting the flux of identified sources, then making another pass to look for weaker sources using a lower threshold. If the threshold for declaring a source is set so low that noise peaks exceed it, then the program happily identifies random clusters of noise as sources as well.

Noise from the receiver and imaging process generates some surprisingly bright spots, so how do you tell real signal from noise peaks? Statistics says that in n averaged samples, the largest peak due to noise should be about $\log(n)$ times the variation about an average value. Images for each channel consisted of 1024 x 1024 = $\sim$10^6 pixels ("$\sim$" means approximately or order of magnitude), and each synthesized beam covered about ten pixels, so there were about 10^5 independent places where either noise or real flux might be in an image. The biggest noise peak in one image would be $\log(10^5)$ = 5 times the noise, and the biggest in a 127-channel cube would be about seven times.

After hours of calculation, SAD found all continuum sources (already known from channel Ø images), some possible weak narrowband sources, and lots of noise peaks. The features it found after ignoring the continuum sources are shown in Figure 5.7; a few peaks were stronger than six times the noise level, but none were higher than the seven sigma peak value that could come from noise alone.

I investigated many of the narrowband features—scrutinizing their spectra, combing catalogs for stars at the same positions, and massaging the data

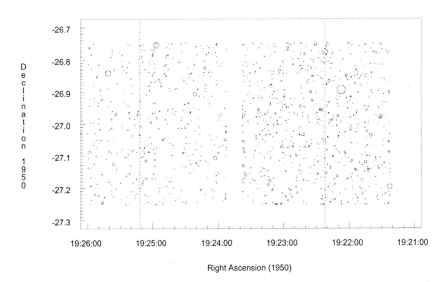

Figure 5.7. Spectral features stronger than 4.2 sigma in 127 channels all superimposed, for the two fields containing the Ohio State coordinates, with continuum sources removed. The Wow right ascensions are shown by dashed vertical lines, and the features are shown by circles scaled to flux. Most and probably all of the features are due to noise.

to tease out weak signals. Nothing exciting emerged; all I had harvested seemed to be noise.

PROVING NOT LITTLE GREEN MEN

I *had* found radio sources near the Wow positions, and although they looked like natural sources, I needed to prove they were natural before dismissing them. The band of frequency covered was only one-tenth of the width of a television signal, so it was conceivable that one of the seemingly natural sources might be ET sending us *I Love Lucy* or something else entertaining using a wide-band transmission.

Pat Palmer suggested looking at a much higher frequency because the sources I'd found should appear there as well, if they really were continuum sources; if they didn't, then they would be worth looking into as possibly artificial. There were other reasons to return to the telescope. One was to look longer to improve the chances of catching intermittent signals, and looking longer would increase sensitivity to weak sources. Another reason was to search a wider band, in case the Wow wandered in frequency. I dashed off a proposal for four more hours and again got approved.

I returned to Socorro in May, 1996, a veteran confident in my ability to run the Great Machine. I napped for a few hours in the guesthouse, then drove out into the night at two in the morning for my three o'clock observing run. Somehow I missed a turn and found myself lost on an uphill gravel road staring at a sign in my headlights stating DUSTY—presumably the settlement I'd visit if I kept driving further into the mountains in the dark. I backtracked fast and arrived at the observatory late.

My observing program had started automatically, and the operator had intervened to correct a mistake in my instructions. Humbled, I watched the spectrum analyzer for the next few hours while the great antennas tracked across the sky in the dark. Several big spikes surged up out of the noise, but while the array was pointing away from the Wow position, and they faded when the array turned back—probably local interference.

The array gazed at each of the dozen sources previously detected, listening for a few minutes at 4850 MHz to see if they glowed there too. It also listened for 43 minutes on each of the two Wow locales at the hydrogen frequency, covering a band 1.5 MHz wide where a drifting signal might lurk. With channels twice as wide as before and listening for twice as long, the observations were twice as sensitive (sensitivity increasing with the square root of both time and bandwidth $\sqrt{2} \times \sqrt{2} = 2$).

When I analyzed the data, all of the previous sources were detected at the higher frequency as well, which confirmed them as natural, and no narrowband features were much stronger than what was expected for noise peaks . The threshold this time was roughly 5 sigma, lower than earlier observations because the array was in a more compact "D" configuration making synthesized beams larger and images only 256 pixels on each side.

Finding no sign of signals, I had to conclude that the "dim but constant" idea was a bust.

SUMMARY

The VLA found no sign of a signal down to almost one-thousandth of the Wow's strength, which was not entirely bad news—it ruled out two more possibilities. First, it showed that there is probably no weak underlying beacon that's "on" all the time and just required a big antenna to reveal, although a beacon could always be weaker than we can detect. Second, it showed that there is no artificial or natural source whose intensity was varying due to interstellar scintillation, rather like the twinkling of stars, because the variation would have to be a factor of 100 times or greater. Scintillation can make things vary in strength by a factor of two and occasionally more, but not by such a large factor.

Only a few possible origins for the Wow seemed to remain, other than interference: either it was an intermittent signal illuminating us rarely, or it was drifting in frequency. Signals changing in frequency did not seem to make much sense because they would be harder to find—and much harder to verify—than signals with a constant frequency, so that possibility did not seem worth pursuing. Looking for signals that are intermittent is hard because they might not shine our way very often, but I could imagine many reasons why signals might be intermittent. To look for such signals I would need to dwell for many hours or even days waiting for a flash. The VLA was not very well suited to that purpose and competition for observing time is fierce, so I had to look elsewhere.

CHAPTER 6

SETI DOWN UNDER

HYPOTHESIS: LIGHTHOUSE?

One of the creative arts in science is imagining how things might have come to be, sometimes spinning theories from few threads of fact. This chapter describes the best scenario I could weave to explain both the original detection of the Wow! and why it later seemed so elusive, which led to searching from Australia.

What might make a signal appear intermittently? One simple explanation would be a rotating source, illuminating observers occasionally like a lighthouse sweeping its beam around the horizon. That could explain why attempts to re-detect the Wow failed—the lighthouse might not have been pointing our way when we looked. Since a rotating source should reappear periodically, the obvious strategy for finding it would be to look long enough to glimpse it at least once and preferably repeatedly.

PERIODIC SOURCES

If the Wow was a periodic source, how long might it take to reappear?

The original Ohio detection can be viewed as one success in 18 trials because Ohio's antenna was usually left at the same position for three days, then moved up or down a half of a beamwidth, yielding nine consecutive days that it might detect a strong source. And there were two chances each day because of the two beams, for a total of 18 chances. The probability of exactly one success in that many trials is shown in figure 6.1 for various periods, calculated using the binomial distribution (same as for flipping coins).

The chances of Ohio's search snagging a signal with a period longer than about 12 hours worked out to less than 10%, and under 5% for 18 hours, which suggests that listening that long might be enough to catch a periodic Wow—unless Ohio had been awfully lucky and happened to catch something repeating even more rarely.

Intriguingly, periods that long include planetary days. Jupiter and Saturn have days about 10 hours long, Uranus and Neptune about 16 hours, and the Earth and Mars are about 24 and 25 hours, respectively. That suggested one possible scenario: a broadcast from a stationary antenna on the surface of a planet, sweeping across us once each extraterrestrial "day."

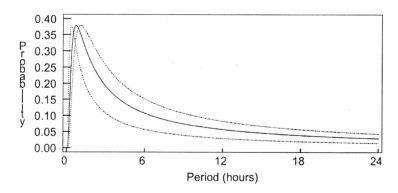

Figure 6.1. Probability of detecting a periodic signal exactly once in 18 trials. The middle line represents a signal lasting 144 seconds and repeating with a period shown on the horizontal axis. The upper line represents a signal duration twice as long, and the lower line half as long. The probability of exactly one detection falls off for periods less than about one hour because multiple detections are not counted. Source: Gray & Ellingsen, *Astrophysical Journal* 2002, reproduced by permission of the AAS.

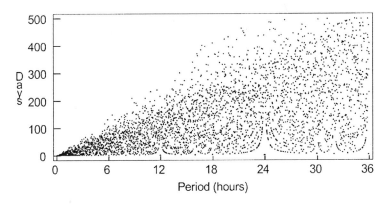

Figure 6.2. Number of consecutive days (vertical scale) until a second opportunity for detecting a periodic signal with the Ohio State transit telescope for sources with various periods (horizontal scale). Chances for detection are especially poor for sources with periods near but not exactly the same as multiples of the terrestrial 24-hour day, and for longer periods. Source: Gray & Ellingsen, *Astrophysical Journal* 2002, reproduced by permission of the AAS.

A rotating source might explain why Ohio State failed to get any more detections during their one hundred subsequent brief observations—after the initial detection, the rotating Earth and rotating source would need to synchronize before the receiver and transmitter would point at each other again. Computer simulations showed that for nearly half of the periods assumed for a source, Ohio would not get a second detection for a hundred days or more, as illustrated in Figure 6.2. Ohio's observations would have been a poor test for signals that were periodic due to rotation or any other cause.

Tracking the Wow locale for something like 18 hours seemed like a good way to catch a periodic signal, but the position can't be tracked for more than four or five hours daily from the Northern Hemisphere because it's below the horizon most of the time like the winter Sun. But it's above the horizon longer at southern latitudes, so I proposed a search at the 210-foot antenna at Parkes, Australia, which could track the locale for nine hours. My first proposal was rejected, with the reviewer suggesting that the signal was due to an American spy satellite. A second proposal citing reasons to doubt a satellite origin was also declined, even though a prominent local astrophysicist named Ray Norris joined in the proposal. Ray suggested contacting his former student Simon Ellingsen, a post-doctoral research fellow who was running a 26-meter (85-foot) radio telescope outside Hobart, Tasmania. I wasn't quite sure where Tasmania was.

MT. PLEASANT

Tasmania is an island state of Australia located a few hundred miles south of the mainland. The capitol city Hobart is home to 200,000 souls and is known among other things as the destination of the annual Sydney to Hobart yacht race. Hobart hosts the University of Tasmania, which operates a radio observatory at Mt. Pleasant about 15 kilometers northeast of town, shown in Figure 6.3.

Simon was searching for methanol masers and had discovered many—bright blobs of alcohol molecules in the disks around newly-formed stars, at just the right distance to be excited into laser-like action. It was a hot topic in astrophysics and he was not terribly interested in SETI, but the Wow was intriguing enough to divert some of his attention and telescope time.

The Mt. Pleasant telescope was especially well-suited to hunt the Wow because it could track the locale for almost 14 hours, long enough to have a good chance of seeing an occasional flash. Its receiver system had 512 channels for each of two circular polarizations, each channel 5 kHz wide, roughly comparable to Ohio's and covering a much larger band of frequency. The system was not as sensitive because the antenna was smaller, but we calculated that by integrating for 30 seconds, the Wow would stand four times higher than the noise peaks, which would be good enough.

Figure 6.3. The 26-meter antenna at the University of Tasmania's Mt. Pleasant Observatory. Courtesy of Simon Ellingsen.

HYDROGEN

Simon tracked the two Wow positions in October, 1998, for 14 hours each and I copied the data from his computer in Tasmania to mine in Chicago via the Internet, reduced it into usable form, and ran it through a battery of analysis programs. A sample of what we saw is shown in Figure 6.4. The most prominent feature was hydrogen, spanning many channels because the gas moves with a range of speeds. Several obvious radio signals were seen but remained when the antenna was pointed off-source—a sign of local interference—so we ignored those channels.

One interesting finding was that hydrogen emission peaked at 15 to 20 kHz below the frequency it would have if it was motionless in the local standard of rest, shown in Figure 6.5. The Wow frequency given by Kraus was also about 20 kHz below that frequency, which suggests but hardly proves that the Wow might have been broadcast at the hydrogen frequency in the same distant place as the hydrogen we detected, both being Doppler shifted similarly due to local motions at the source.

There was, however, no sign of a signal like the Wow, much less anything popping up periodically. Why not? The notion of a periodic signal might be

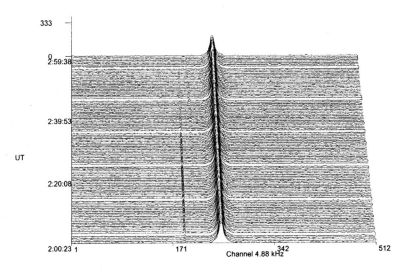

Figure 6.4. An hour of Hobart spectra at 30-second intervals showing the hydrogen line in the center and a local radio signal to the left of that. The horizontal axis is channel number, height is flux in janskys, and successive spectra are stacked so time (UT) is on a vertical scale as well.

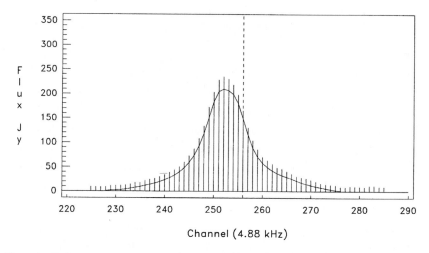

Figure 6.5. Hydrogen spectrum averaged over several hours (curve), with the biggest peak during any observation (needles). The dashed vertical line shows the channel where hydrogen would fall if motionless in the local standard of rest; its observed frequency is 20 kHz lower because it is moving away from us at three or four kilometers per second.

wrong, but it seemed like too good an idea to throw out hastily (and we didn't have any better ones). Perhaps we had not looked long enough—maybe the lighthouse turns slowly, like the Earth's 24-hour rotation or longer.

It was also possible that we did not look in quite the right spot. Ohio's beam was a Moon and a half tall and could have picked up sources located above or below the area covered by the Moon-size Hobart beam, so we decided to look north and south of the Ohio positions. It was also possible that a signal was hidden beneath the hydrogen emission, so that waiting roughly six months for Doppler shifts to change as the Earth moved through its orbit would shift the received frequency of hydrogen and uncover a buried signal.

In March of 1999, Simon observed a quarter-degree north and south of the original positions, and again my analysis found no obvious sign of the Wow. One feature was notable—with a flux of about 60 janskys and occupying two 5-kHz channels (shown in Figure 6.6), just like the Ohio State signal. The signal did not look enough like the Wow to consider it a re-detection, however, because it persisted for only one 30-second observation and was not near the Wow frequency. But it merited a second look. In November of 2005, Simon looked at that spot again for seven hours and found much more inter-

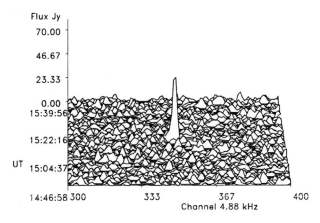

Figure 6.6. Feature in channels 349 and 350 during one 30-second observation north of the 19^h 22^m location.

ference than before but no sign of the feature. Growing interference ended the usefulness of Hobart for hunting the Wow.

SUMMARY

We had a good chance of detecting signals repeating every 14 hours or more often, so it seems pretty certain that the Wow was not a periodic signal flashing more often than that. The chances of detecting signals with longer periods were roughly 80% for periods up to 24 hours, although a 14-hour observing run would have no chance of detecting a source with a longer period if the run happened to start just after an emission cycle ended.

We have not pressed on with more observations at Hobart because of increased interference and the large amount of telescope time needed to monitor a half-dozen positions for several days each. If the Wow was real, it should be there to find in the future as newer and more powerful radio telescopes become available.

WOW EPILOGUE

So, *was* the Wow! an interstellar signal?

I must admit that it might have been some sort of man-made interference, since it has proven so elusive, but many things about the Wow suggest a signal from the stars. The evidence seems strong enough to justify looking harder because the Wow is the *best* candidate signal ever seen.

It's possible that a signal is lurking out there and just requires listening longer; all straight-on observations add up to less than one day at any one spot. No search would have been sure to catch a beacon sweeping across our skies every 24 hours, for example, like a lighthouse-style broadcast from a planet with a day the length of our own. A better test of the lighthouse idea would be to look longer—long enough to catch signals repeating every 50 or even 100 hours, which should cover the lengths of a "day" for many planets that might be the source. Looking long is inconvenient but might be what's required to catch some kinds of signals.

The lighthouse notion might, on the other hand, be plumb wrong. The next-best idea might be a signal sweeping across the radio spectrum every so often, which suggests listening for an extended period of time for the signal to return—the same search strategy as looking for a lighthouse—or perhaps scanning the radio spectrum hoping to catch a moving target.

Other searches have reported candidate signals, but none as intriguing as the Wow. Harvard's long-running META survey, for example, reported unexpectedly strong peaks at 37 different positions in the sky, but consisting of only single brief measurements that might have been due to glitches in the computer system. The Wow was seen in *six* separate measurements and they had the signature of a celestial source passing through one of the telescope's beams, which makes it a much stronger candidate—clearly a real radio signal and very likely celestial.

One strange twist in the Wow story is that nobody has tried to independently confirm it, other than myself and my collaborators. There are no reports of other searches in the literature, and nobody in the field that I've talked to knows of any other searches, although many people have the impression that *everybody else* has looked for it. Some people have the opinion that the Wow is not a good candidate, and therefore have not looked for it.

Signals like the Wow may be telling us that somebody or something is out there transmitting, but in ways that our current search strategies aren't good at confirming. We may need to follow up on many leads like the Wow before we finally find anything—but if we do find anybody out there, it's likely to prove interesting.

PART II

Searching for Extraterrestrial Intelligence

CHAPTER 7

OTHER WORLDS

INTRODUCTION

The search for extraterrestrial intelligence, or SETI, looks for signs of life on other worlds from the comfort of our own. Few signs would reach across the enormous void between stars, but radio or other electromagnetic signals are one sign that we know could be detected. Looking for signals limits us to finding only intelligent, technological life—gizmo-builders—but they might be the most interesting kind of critter out there. With radio and optical telescopes, we can search many stars quickly and fairly inexpensively—each one a place that would take titanic resources and generations to reach by space travel, which won't be an option real soon, anyway.

Looking for signals from other worlds makes sense *if* some other stars host planets, and *if* life has emerged on some of them, and *if* in some cases intelligence has evolved, and *if* "they" use stuff like radio. Until recently, this was all speculative—an awful lot of big *ifs*. But the recent discovery of planets orbiting many other stars put firmer ground under such speculation because it means that habitable planets probably exist by the billions. Whether life does stir or think or broadcast anywhere else is still iffy, but potential homes seem to be available everywhere we look out into space.

Finding others from afar would be easiest if they are broadcasting to the stars, but if they aren't, we might be able to eavesdrop on local radio use like radar—we don't need to believe aliens are trying to talk to us. We don't need to believe there's anybody out there at all. SETI merely aims to find out, just one facet of exploring the cosmos around us.

EARLY THOUGHTS ON THE MULTIPLICITY OF WORLDS

The idea that other worlds exist goes back thousands of years. The Greek philosopher Epicurus wrote in the fourth century BC in his *Letter to Herodotus*:

There are infinite worlds both like and unlike this world of ours... We must believe that in all worlds there are living creatures and plants and other things we see in this world.

In about 70 BC, the Roman poet and philosopher Lucretius wrote in his poem *De Rerum Natura* (On the Nature of the Universe):

For there is such a huge supply of atoms that all of eternity would not be enough to count them; there is the force which drives the atoms into various places just as they have been driven together in this world. So we must realize that there are other worlds in other parts of the universe, with races of different men and different animals.

Giordano Bruno, a Dominican monk, wrote in 1584 in his *On the Infinite Universe and Worlds* that there was an infinite number other suns and worlds like the Earth. His views on theology put him at odds with religious authority, and he was imprisoned, examined, excommunicated, and burned alive in 1600 for heresy by the Inquisition. Some scholars believe that the charges against Bruno focused more on religious doctrine than his belief in the plurality of worlds.

Early ideas about other worlds were speculation because no other worlds anything like ours were known to exist. The Sun and other heavenly bodies were thought to be made of quite different stuff than Earth, and most people believed that those bodies revolved around the flat and motionless place they lived—if they thought about it at all. The emergence of modern science changed that world view.

Nicolaus Copernicus advanced the heliocentric theory in 1543—that the Earth and planets circle the Sun. The idea flew against common sense of the time; there was no obvious reason to believe that the Earth moved or that it was not the center of things. Johannes Kepler figured out that the orbits were ellipses rather than circles in 1605 (the elliptical shape is often exaggerated in drawings; Earth's orbit differs from a circle by less than two percent). Galileo Galilei turned the newly invented telescope on the skies in late 1609 and early 1610 and discovered mountains on the Moon, four moons orbiting Jupiter, and the phases of Venus—showing that they were real places and not merely points of light in the heavens. Galileo also fell afoul of the Inquisition for advocating the heliocentric view in *Dialogue Concerning the Two Chief World Systems*, with Pope Urban VIII's favorite theory argued by a character named Simplicius which suggested simple-minded. He spent his old age under house arrest, probably spared worse by an earlier friendship with the pope.

Isaac Newton finally explained the movement of the planets around the Sun, using the idea of gravitational attraction and a few simple principles of motion. By the time he published *Philosophiæ Naturalis Principia Mathematica* (Mathematical Principles of Natural Philosophy) in 1687, it was clear that observations, experiments, and mathematics were better ways to understand the physical world than authority and dogma.

By the 1800s, the idea that intelligent creatures might exist elsewhere in our solar system was downright popular. The mathematician Karl Friedrich Gauss, for example, proposed revealing ourselves to the inhabitants of the Moon by signaling with mirrors or planting wide strips of trees in geometric shapes, and on August 25 of 1835 the *New York Sun* newspaper began printing a series of well-received (and false) reports on telescopic observations of intelligent Lunar creatures. As the twentieth century neared, popular books and newspaper stories depicted a race of intelligent canal builders on Mars, based in part on astronomer Percival Lowell's mistaken interpretation of fuzzy features on the planet as an irrigation system (the Italian astronomer Giovanni Schiaparelli had described them as "canali", meaning channels, but not artificial). In 1900, the Guzman Prize of 100,000 francs was offered by the French Académie des Sciences for the first communication with the beings of another world *other* than Mars, which was considered too easy; the prize was eventually awarded to the crew of Apollo 11.

The twentieth century brought enormous advances in astronomy, so that today notions of other worlds are based on solid evidence. Humans have walked on the Moon and brought back pieces of it, landed spacecraft on Mars, Venus, and other bodies, and have looked down on much of the solar system from spacecraft cameras. We now know that planets orbit other stars and have good reasons to think that many exist and that some are habitable.

STAR BIRTH

Stars and planets form at the same time out of vast clouds of interstellar gas and dust like the one shown in Figure 7.1—roughly two-thirds hydrogen, one-third helium, and a dash of other stuff. Gravitational attraction slowly pulls

Figure 7.1. Stars forming in the Eagle nebula M16. Light from most of the young stars is hidden by the dust, although their heat can be seen in infrared images. Photo credit: Jeff Hester and Paul Scowen (Arizona State University) and NASA/ESA.

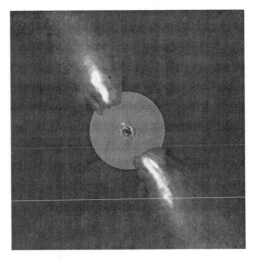

Figure 7.2. A circumstellar disk orbiting the star Beta Pictoris, viewed almost edge-on in infrared light. The disk, composed of dust and ice grains, was observed in 1996 using the European Southern Observatory 3.6 meter telescope. The inner part is an image from the ESO's Very Large Telescope with the central star and parts of the disk subtracted to reveal what appears to be a planet just upper left of center. The star is about 12 million years old and 60 light years away. Source: ESO/A.-M. Lagrange *et al.*

some of the gas and dust together, and the cloud begins rotating as it contracts like a figure skater spinning faster when outstretched arms are pulled in. It flattens into a disk with a huge ball of gas at the center, and gravity compresses the core to a density more than ten times lead, heating it enough to ignite nuclear fusion. That fusion process transforms hydrogen atoms into helium, releasing energy radiated partly as light, and a star is born.

The rotating cloud flattens into a circumstellar disk—somewhat like the rings of Saturn but on a far grander scale, as shown in Figure 7.2. Dust grains collide and sometimes stick together aggregating over time into planetesimals the size of hills or mountains. Planetesimals grow into planets by mutual gravitational attraction and collisions over millions of years. Leftover gas, dust, and chunks are blown away by solar wind when the star begins shining, swept out beyond the planetary system, some of the debris to return later as comets.

This so-called nebular hypothesis was sketched out by Immanuel Kant in 1755 and in more detail by Pierre-Simon Laplace in 1796, based on the observation that the planets all orbited the Sun in the same plane and in the same direction, and the Sun rotated in the same direction too. But a competing theory had it that our planets were formed from material ripped from the Sun by a passing star—and that implied planets would be scarce around other stars because stars rarely cross paths. The rival theory held sway for a long time because the Laplace-Kant picture had a big problem—it predicted that the Sun forming from the same swirl of stuff as the planets would spin faster than it does. That seemingly fatal flaw was removed only in the mid-1900s, when it was realized that the Sun's rotation has been slowed by its magnetic field.

If planets are spin-offs from star birth, then many or most stars should harbor planetary systems. But until 1995, none had been found despite many attempts and occasional mistaken claims of discovery.

EVIDENCE OF OTHER WORLDS

Confirmed claims of planets orbiting other stars were made in 1992 by Aleksander Wolszczan and Dale Frail, who detected several orbiting a supernova remnant called PSR 1257+12, and in 1995 by Michel Mayor and Didier Queloz, who detected a planet orbiting the normal star 51 Pegasi. Within the next few years, many more were found by the team of Paul Butler and Geoff Marcy as well as others, and in 1999 the first *system* of planets was discovered orbiting a nearby star. Within a decade, hundreds of extrasolar planets had been found—a new world every few weeks, an utterly unprecedented pace of discovery. By 2012, thousands of candidate planets had been found by the Kepler spacecraft and more were expected.

Planets must be common because the first few hundred were found by looking at only a few thousand nearby stars. There must be billions of them, which is a very different view of the universe than the one where only eight or nine planets were known to exist.

Evidence for many of the first extrasolar planets found came from tiny shifts in the color of starlight. Just as a star's gravity tugs on planets to form their orbits, planets also tug on their star, causing a reflex motion—movement of the star in a tiny orbit caused by the planet. Jupiter, for example, makes the Sun move in a little circle one-thousandth as big as Jupiter's orbit (the planet's mass is one-thousandth of the Sun's), and that makes the Sun move at 13 meters per second as the giant planet swings around it once every

12 years. A distant observer seeing our star moving to and fro could deduce the presence of Jupiter without actually seeing it, and could estimate its mass and length of its year.

Reflex motion is detected by measuring the wavelength of spectral lines in starlight and looking for Doppler shifts. When a star is moving away from us in its gravitational dance with a planet, its light reaches us at a slightly lower frequency than usual and spectral lines are very slightly shifted toward the red end of the spectrum, or "red shifted". When a star is moving toward us, the lines are shifted toward the blue end. Motion of one meter per second—a casual stroll—can be detected over enormous distances, which is an astonishing achievement. An example is shown in Figure 7.3.

The first few hundred extrasolar planets did not seem much like those in our solar system; many were more massive than Jupiter and orbited their stars very closely. It's no surprise that smaller, Earth-like planets seemed scarce because the reflex method is most sensitive to massive planets orbiting close in. The masses of the Sun, Jupiter, and Earth compare like 1000, 1, and 1/318—very roughly a pumpkin, a plum, and a pea. With only a small fraction of Jupiter's mass, the Earth tugs on the Sun only about 0.1 meter per second—an infant's crawl—and that can't currently be detected. It's also no surprise that many extrasolar planets orbit their stars close in with periods as short as

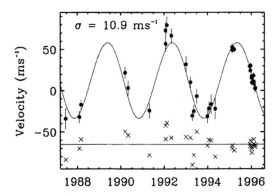

Figure 7.3. Reflex motion of the star 47 Ursae Majoris, 50 light years away. The cyclic variation in its velocity over the years, shown along the horizontal axis, indicates a planet orbiting the star once every three years, with a mass several times that of Jupiter. The nearly sinusoidal shape of the curve fitted to dots representing velocity measurements reveals a nearly circular orbit. Source: Butler & Marcy, *Astrophysical Journal* 1996, reproduced by permission of the AAS.

a few days, because we have not yet looked long enough to accumulate the several orbital cycles needed to detect planets in more distant orbits.

Planets can also be detected by a transit method if the planet happens to pass in front of its star, making the star's brightness dip a bit. The dip is small—only one-hundredth of one percent for a planet like the Earth in a similar orbit, but detectable by telescopes above our shimmering atmosphere. The amount of the dip tells us the planet's size, and recurrences give the length of its year. France's COROT space mission and NASA's Kepler mission both detected candidate extrasolar planets using the method, Kepler generating over two thousand by the end of 2011, some illustrated in Figure 7.4. Many candidates were near the smaller end of the size scale, suggesting that Earth-size planets might be common.

Some newfound planets do seem to echo the pattern of our solar system. Systems of planets have been found, with worlds of various sizes, and some are roughly the size and distance of Jupiter and Saturn. Some planets compa-

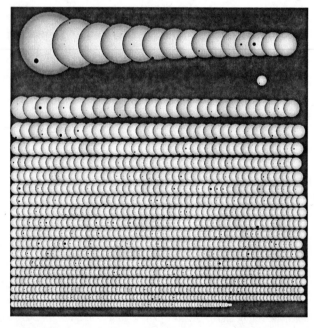

Figure 7.4. Illustration of 1,235 planet candidates transiting 997 stars, from the Kepler spacecraft as of early 2011. The large circles represent the size of stars with transiting planets, which in turn are represented by small dark spots, all scaled to the lone circle and dark spot representing the Sun with Jupiter transiting. Credit: NASA, illustration by Jason Rowe.

rable in size to the Earth have been found, most orbiting so close to their stars that they would be too hot for life, but a few in habitable zones.

Now, it is certain that there are other worlds, and probably billions of them, because a lot have been found by scrutinizing only a small sample of stars. Planet searches are finding that between 25 and 50% of Sun-like stars have planets—after adjusting for limitations of instruments and methods and not having looked very long—and it's possible that essentially all such stars have planets. One early prediction by Wesley Traub is that one-third of Sun-like stars may have terrestrial planets that are in habitable zones, where life might be possible. Approximately 10% of all stars are Sun-like (spectral types F, G, and K), so several percent of all stars may harbor planets.

The sky is not just full of stars, but full of other worlds.

MYRIAD STARS; MYRIAD PLANETS

If anything like one percent of the myriad stars have planets, then an enormous number of planets exist ("myriad" means both ten thousand and countless in Greek; stars really are countless, since new ones are forming all the time). Planets and therefore the possibility of life seem to exist everywhere you look out into space.

The human eye can see the light of only about five thousand stars through the small opening of its pupil, but telescopes collecting light over a larger area reveal vastly more stars too faint to see with the unaided eye. Our Milky Way galaxy contains several hundred billion. To illustrate how many that is, imagine that stars are the size of coarse grains of sand, about one millimeter across. A million of them in a single layer would cover one square meter—a smallish window, for example. A hundred million would cover the walls, floor, and ceiling of a room—everywhere you looked, you would see one. A hundred billion grains of sand would fill the room from floor to ceiling!

Our galaxy is a vast whirlpool of stars thought to look like the examples in Figure 7.5. It's about one thousand light years thick in our neighborhood, and about one hundred thousand light years across. A light year is the distance light travels in a year: six trillion miles, ten trillion kilometers, or 10^{16} meters; the nearest star is four light years away. The Sun is somewhat more than half-

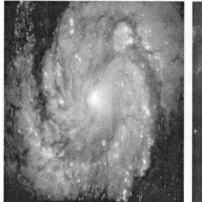

Figure 7.5. The Milky Way is known to be a spiral galaxy, and thought to look like M100 (left) or NGC 1300 (right) if viewed from above or below. Credit: NASA/ESA (M100); NASA, ESA, and The Hubble Heritage Team STScI/AURA (NGC 1300).

way out from the center and near the plane of the disk; it orbits the center once every several hundred million years. We see the Milky Way as a patchy band of light across the night sky because the disk surrounds us; it would look brighter if most of its stars were not obscured by dust and distance.

With so many stars in our galaxy, a simple calculation suggests there could be a thousand radio-using civilizations if only one percent (10^{-2}) of its 10^{11} stars has a habitable planet, and only one percent of those planets harbor life, and only one percent of those also has intelligence, and only one percent of the intelligent critters use radio ($10^{11} \times 10^{-2} \times 10^{-2} \times 10^{-2} \times 10^{-2} = 10^{3}$). If only one percent of them broadcast strong enough signals, there might be ten in the galaxy that fairly modest efforts might detect. "One percent" is of course arbitrary; we don't know how often habitable planets, life, and intelligence might occur, or how long broadcasting might last.

Looking further out into space, the number of stars soars. Our sister galaxy Andromeda (also called M31, from its number in Charles Messier's *Catalog of Nebulae and Star Clusters* published in the late 1700s) contains roughly a thousand billion, or 10^{12}. It's barely visible to the naked eye or binoculars as a smudge on the sky six moons wide, a few million light years away. Several dozen additional smaller galaxies comprise our local group, and the observable universe is thought to contain roughly a hundred billion other galaxies of various shapes and sizes—perhaps 10^{22} stars in all, more than the number of sand grains on and below all of our beaches.

Most of the universe seems too distant for us to detect life with known methods, but it seems feasible to detect radio or optical signals from millions of stars in our neighborhood of the Milky Way, and perhaps from many billions of stars throughout our galaxy and others in the local group.

If a large number of other worlds exist, there is probably a great deal of variation among them. Many are surely lifeless—too hot or too cold, too harsh, or not made of the right stuff. But some probably have Earth-like temperatures and similar raw materials, so if life arises readily or even just occasionally, many planets might teem with life. So far, however, we know of only one world where life exists, so "Does life exist elsewhere?" is an open question.

Chapter 8

Life

Life on Earth

Life stirred on Earth quite soon after the planet formed, and has survived every adversity for billions of years—near-boiling temperatures, ice ages, asteroid and comet strikes, changes in the very air, and much more outrageous fortune. Arising so fast and surviving so long, many scientists think life might also exist elsewhere.

Our solar system is thought to have been born about 4.5 billion years ago, when a vast cloud of interstellar gas and dust slowly drew together into a swirling disk to form the Sun and planets. The early Earth was bombarded by asteroids and comets left over from star birth and was probably molten from the heat of impacts and its own internal radioactive decay. One titanic collision with something the size of Mars is thought to have splashed off enough stuff to form the Moon. The bombardment gradually subsided as growing planets swept up much of the space junk, and after a few hundred million years the surface cooled enough for water to condense. An early atmosphere belched up from volcanoes and rained down from comets, but that early "air" would kill you quick because it had no oxygen. It seems that life began between 3.5 and 4 billion years ago, remarkably soon after the planet's hellish birth (the official name for that geologic eon is Hadean).

Some scientists have suggested that life first formed near the surface of the sea or in smaller bodies of water, in a "warm primordial soup" of chemical building blocks brewed by sunlight and lightning. Chemists have shown that simple organic compounds, containing carbon and found in all living things, could have been formed by natural processes. In 1955, for example, Stanley Miller and Harold Urey produced many of the amino acids that are basic to life by sparking a mixture of gases in a flask to simulate the effect of lightning on an early atmosphere of hydrogen, methane, ammonia, and water vapor. It's now thought that the early atmosphere was mostly carbon dioxide and nitrogen with a little hydrogen and ammonia, a recipe that yields organic molecules in experiments too, although not so many.

The discovery of organisms living around volcanic vents on the ocean floor raises the possibility that genesis occurred deep under water and without sunlight. These so-called black smokers provide energy to warm the seawater, plus minerals and gases rich in carbon, hydrogen, and other stuff. One advantage of a deep-sea womb is that it would insulate emerging life from the harsh early surface—ultraviolet light, asteroid and comet impacts, volcanic eruptions, and other annoyances. Some amino acids and other complex molecules have been produced in laboratories simulating conditions near deep sea vents.

Some of life's raw materials may have arrived as meteors and interplanetary dust. Radio astronomers have found over 100 kinds of molecules drifting in space including organic compounds, and tons of space dust and meteorites rain down onto the Earth each day. Asteroids and comets are thought to have delivered some of our water.

Some raw materials may have been cooked up in volcanoes, including some of our water, boiled out of rock. The ingredients for life may have come from many or all of these sources.

Exactly how these molecules got together in the complex chains and systems that live—metabolize, reproduce, and evolve—is still a big mystery. The earliest "living" thing was probably just a molecule that could make copies of itself, a simple prototype for the genetic material found in all living things today. It seems wildly improbable that random molecular jitter alone would conjure up such things even over ages, and ultraviolet light or heat would tend to break the molecules apart. Some think that a mineral like clay or pyrite might have served as a catalyst to hold small molecules long enough for more complex ones to form, perhaps sheltering them from damaging light.

An alternative explanation for life on Earth is panspermia—the notion that self-replicating molecules or simple organisms from somewhere else somehow arrived and thrived on the Earth. It doesn't explain how life began, but it allows more time for it to happen; instead of being born soon after the Earth's formation, life would have billions of years longer to emerge elsewhere. Panspermia envisions spores of life spreading from planet to planet— perhaps blasted off by meteor strikes and sailing from star to star on solar winds, or even spread by interstellar travelers (tossed out with the garbage, in the cartoon version). It's not known how life could survive the harsh conditions of space and spread across the vast distances between stars, but one notion is that microorganisms could ride inside rocks—frozen, dormant, and protected from radiation during the voyage, then shielded from the blazing entry through a planet's atmosphere.

Divine creation is yet another way to account for life. Most but not all scientists think life arose through natural processes, but if it was the act of a Creator, it should be possible to create life on other planets as well. Whatever or whoever gave rise to life on Earth, there is no obvious reason why the same thing couldn't happen elsewhere.

A BRIEF HISTORY OF LIFE

Fossil evidence of life goes back at least 3.5 billion years, to the first quarter of the planet's history, as illustrated in Figure 8.1. One kind of fossil that dates back that far is a mushroom-shaped rock that resembles today's stromatolites—sedimentary structures formed by colonies of bacteria, shown in Figure 8.2. Another kind of evidence is microscopic shapes in rock that are thought to be fossilized bacteria, found near stromatolites. Simpler organisms must have come earlier but left few traces because early history has been erased from most rock by melting, weather, and other ravages of time. There

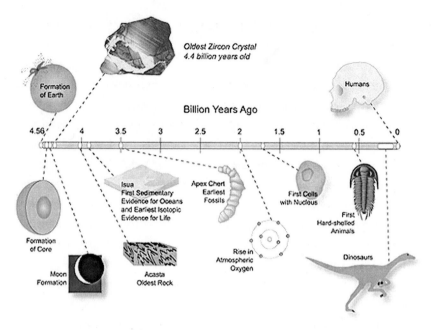

Figure 8.1. Life on Earth timeline, illustrating major events and the relatively recent appearance of large creatures. Source: NASA.

is some evidence of life in rock as old as 3.85 billion years—an enrichment in the carbon-12 isotope that's a signature of biological processes.

For its first two billion years, life consisted of anaerobic microbes—single-celled critters that don't need oxygen—which was lucky for them because there was hardly any in the early atmosphere. They fed on compounds based on things like sulfur and iron, and later cyanobacteria developed the trick of photosynthesis—feeding on sunlight and excreting oxygen. Over time, these little fellows are thought to have built up our oxygen-rich atmosphere, a big job because the reactive gas is constantly being soaked up in chemical reactions like rusting iron. Rising oxygen levels were toxic to anaerobic microbes, but between one and two billion years ago aerobic ones appeared and thrived in the stuff.

Roughly a billion and a half years ago, a new domain of life appeared: Eucarya (from "good nut" in Greek), which had internal structures like a nucleus and specialized organelles like mitochondria. That seems to have made larger multi-celled organisms possible, including animals. Animals use

Figure 8.2. Stromatolites today, on the seashore (left) and in cross section (right). The layers are due to precipitation of calcium carbonate (lime) from seawater and debris being captured in a layer of mucus over mats of cyanobacteria. Source: Cyanosite, www-cyanosite.bio.purdue.edu.

oxygen to fuel a metabolism that produces enough energy to move about, allowing them to feed on higher-energy food—plants and other animals— and to develop complex behavior.

Something over a half-billion years ago, larger creatures appeared during the Cambrian explosion, a period when many new species appear in the fossil record, including the familiar trilobites. The first vertebrates appeared (jawless fish), and 400 million years ago jawed fish (like sharks) appeared, and the first animals ventured out of water onto land. About 300 million years ago, amphibians arose, able to live on land but needing water for their eggs without shells, followed by the reptiles including dinosaurs, laying shelled eggs better adapted to land. Some 200 million years ago, small warm-blooded mammals emerged with faster metabolisms fueling faster brains, and they began flourishing 65 million years back with the extinction of dinosaurs.

The last few million years produced animals with especially good brains (we like to think), which today look back and ask questions like "Could this riot of life break out elsewhere?"

CONDITIONS FOR LIFE

Life exists under a tremendous range of conditions on Earth—in temperatures ranging from freezing to nearly boiling, on the ocean floor under immense pressure, in places so acidic that concrete dissolves, and in many other hostile spots. Some critters get along without oxygen entirely; some don't need sunlight; some live in rock miles underground munching minerals.

Essential conditions for life are thought to include a liquid medium like water, a source of energy, and the right materials.

Water is widely viewed as necessary for life as we know it. On the small scale of a cell, water dissolves and mixes stuff, bringing atoms and molecules close enough for reactions to occur; water also takes part in chemical reactions. On a larger scale, water dissolves materials and transports them, allowing things to mix and move in rivers, lakes, and oceans. A temperature range allowing liquid water is important because many chemical reactions slow or stop below freezing, and ice kills cells by bursting their walls. At high temperatures, biological molecules fall apart and critters cook. Other liquids might work for very different kinds of life, but water is the most common bio-friendly fluid we know—and H_2O is one of the most abundant molecules in the universe; molecular hydrogen H_2 is the most common.

All life on Earth requires energy to drive its chemistry. Sunlight, for example, drives photosynthesis in plants, storing energy in chemical bonds—fueling the plant, and moving up the food chain by way of animals eating plants. Photosynthesis is a fairly complicated chemical trick, though, and did not evolve until several billion years after life arose. Early life used simpler energy sources, including chemical and thermal, and lightless critters can still be found living far underground in rock and deep in the ocean around volcanic vents on the sea floor.

The "right" materials are thought to include carbon, which is an essential ingredient of life on Earth. Terrestrial life is sometimes called carbon-based because many of the molecules involved in living processes have carbon atoms. Even the simplest living things rely on a rich menu of chemical compounds and reactions, and carbon compounds provide many of the needed ingredients. Many scientists think that life elsewhere might also be carbon-based because those atoms can form so many different compounds—more than any other element, more than all other elements combined. Yet carbon is not common everywhere; very little exists on the surface of the Moon, for one example.

An atmosphere seems important, although perhaps not essential. The pressure of the ocean of air above us keeps our water (and blood) from boiling away, and it also makes atmospheric gases permeate water, allowing fish and marine plants to breath dissolved oxygen or carbon dioxide. An atmosphere also helps regulate a planet's temperature, protecting against extreme changes; the airless Moon, in contrast, has temperature swings of hundreds of degrees between day and night. Atmospheres also help protect the surface from ultraviolet radiation, cosmic rays, meteors, and so on. Smaller bodies like the Moon lack enough gravity to keep gases from drifting off into space, while bodies the size of Earth and Venus (and Mars, barely) have enough mass to hold gases like oxygen, nitrogen, and carbon dioxide. It is quite possible to imagine life on airless worlds—perhaps in an ocean protected from the vacuum of space by a shell of ice, or beneath a crust of dirt or stone—but it seems a lot less hospitable.

The Earth is the only place we currently know of that is definitely habitable: a sizable rocky planet at a temperate distance from its star. How likely is that to happen elsewhere?

HABITABLE ZONES

Astronomers have simulated the formation of planetary systems using computers, "creating" stars and planets by applying the laws of physics and chemistry to the properties of dust and gas clouds in space. Remarkably, the simulations often yield planetary systems roughly like our own. Two kinds of planets usually result: terrestrial ones that are relatively small and dense like the Earth, Mercury, Venus, and Mars, fairly close to the star, and much larger gas giants further out, like Jupiter, Saturn, Uranus, and Neptune.

The orbits of some simulated planets fall within habitable zones—a range of distances from a star where water can exist as a liquid, typically tens of millions of kilometers wide. Figure 8.3 shows some of the many planetary systems generated by Stephen Dole, a pioneer in the area. After conjuring

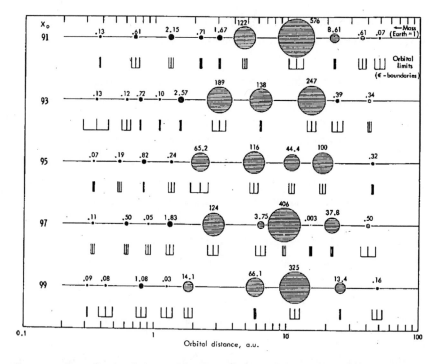

Figure 8.3a. Some simulated planetary systems, with a central star at the far left but not shown. The planets in each system are shown along horizontal lines using a logarithmic scale, with the distance from the star in astronomical units (1 a.u. is the Earth's average distance from the Sun). Circles indicate the relative mass of each planet, with numbers giving their mass in units of Earths. Source: Dole, *Icarus* 1970, used with permission of Elsevier.

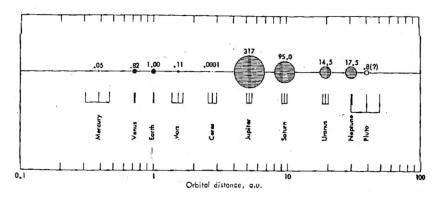

Figure 8.3b. Planets in our solar system. Source: Dole, *Icarus* 1970, used with permission of Elsevier.

up many different systems using a variety of initial conditions, he found that roughly one in a hundred stars would have an Earth-size planet in a habitable zone among its simulated planets.

Habitable zones change over time because stars get hotter as they age, so there's a risk that the band of moderate temperature might sweep past a planet before life has time to evolve very much. The Sun, for example, is 30% brighter now than when the Earth formed and is expected to get brighter over the next few billion years making Earth too toasty. Simulations that include effects of an atmosphere show that habitable zones can remain fairly wide over billions of years because atmospheres help regulate temperature.

Our Sun's habitable zone today seems to begin just inside the Earth's orbit and extends much of the way out to Mars. Mercury is so close to the Sun that its temperature is far too hot to be habitable. The next three planets—Venus, Earth, and Mars—orbit at distances where liquid water depends more on atmosphere and size than on the intensity of solar radiation, and Venus and Mars may have had water in earlier times. The planets further out—Jupiter, Saturn, Uranus, Neptune (and Pluto)—are far too cold for water to be liquid, although some of their moons have internal sources of heat.

VENUS

Venus is in many ways our sister planet. It's nearly the same size as the Earth (12,100 kilometers in diameter compared with 12,756 km), is made of similar rocky stuff, and has an atmosphere. Yet it seems utterly uninhabitable, with a surface temperature hotter than an oven. Nearer the Sun than the Earth (108 vs. 150 million kilometers), sunlight is about twice as intense, but Venus is much hotter than the extra sunshine should make it. The problem is the so-called greenhouse effect—gases in the atmosphere letting sunlight pass through to heat the surface, but absorbing some heat radiated by the surface and preventing it from radiating back into space. The heating is a bit like a greenhouse, although greenhouse glass simply keeps warm air from blowing away. Carbon dioxide is the guilty greenhouse gas on Venus, making up more than ninety percent of the atmosphere; ninety times as dense as our atmosphere, its pressure is like being a half-mile deep in an ocean.

What went wrong on Venus? It's thought that the planet may have had oceans in its early days, much like the Earth. Carbon dioxide belched from volcanoes and dissolved back into the oceans, getting locked up in carbonate rocks like limestone before it could do much harm. Too much water vapor in the atmosphere from the stronger sunlight may have produced a greenhouse effect, raising temperatures high enough to vaporize the oceans. With no seas to soak up carbon dioxide from volcanoes, and the stuff baking out of carbonate rocks on the former sea floor, the atmosphere became thick with it and Venus became a permanent hothouse. Where did the water vapor go? Vapor in the upper atmosphere was broken down into oxygen and hydrogen by sunlight; the hydrogen escaped into space, and the oxygen got bound into rocks, leaving mostly carbon dioxide.

MARS

Mars seems to have had the opposite problem—not enough heat. Sunshine on Mars is less than half as intense as on Earth because of its greater distance from the Sun (228 vs. 150 million kilometers), and temperatures are usually far below freezing. The atmosphere today is too thin for liquid water to last long—it's less than one-hundredth as dense as ours, and under such low pressure water on the surface boils away. But Mars may have been habitable in its youth. Images like those in Figure 8.4 show signs that water flowed on the surface billions of years ago, and perhaps very recently as well although only briefly. Current opinion is that Mars was once warmer and wet—at least occasionally—with a thicker atmosphere that might have allowed life to arise earlier in Martian history.

What happened on Mars? It's thought that the planet is a bit too far from the solar bonfire and a bit too small. Warmed in its youth from internal heat, it probably had water vapor and carbon dioxide as insulating greenhouse gases, but carbon dioxide was constantly stripped from the atmosphere, combining with minerals to form carbonate rocks. On Earth, internal heat drives volcanoes, which recycle carbon dioxide locked in rock back into the atmosphere. Mars, however, is about half the Earth's size (6,792 vs. 12,756 km diameter) and cooled fast; most of its volcanoes died long ago. With little carbon

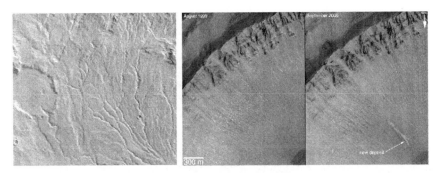

Figure 8.4. Left: Channels on the surface of Mars, evidence of running water in the past although no water is detectable on the surface today. Image courtesy of Lunar and Planetary Institute. Right: Signs of recent surface water on Mars in a pair of images from the Mars Orbiter Camera. The "gully deposit" seen in 2005 (far right) in a crater in the Centauri Montes region was not seen in an earlier image from 1999 (middle). It is thought to be due to the flow of a fluid, probably water evaporating fast. Source: NASA.

dioxide recycled back into the atmosphere for insulation, and less heat from the Sun, Mars froze. Although the atmosphere is ninety-five percent carbon dioxide, it's too thin to have much greenhouse effect.

Today the Red Planet has no visible surface water and very little vapor in its atmosphere; some water is frozen along with carbon dioxide in the polar caps and some is thought to be frozen beneath the surface. The evidence for liquid water on the surface in the past is very strong, and evidence is growing for occasional small flows from underground occurring today.

LIFE ON MARS?

Evidence that life may have once existed on Mars was reported in 1996 by a group of scientists led by David McKay, based on a meteorite named ALH84001 found in the Allan Hills of Antarctica. It's generally agreed that it is a chunk of Martian rock kicked off the planet by an impact and later swept up by the Earth; over a dozen such meteorites have been found, linked to Mars by chemical analysis.

Mineral and chemical deposits that might have been formed by bacteria were found inside the meteorite, along with features resembling fossilized bacteria, shown in Figure 8.5. The claim that the meteorite contained evidence of Martian life was cautious, noting that no single finding was con-

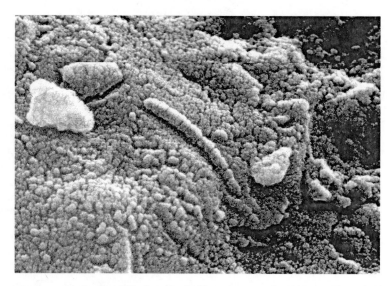

Figure 8.5. A possible fossilized "Martian microbe" from the meteorite ALH84001. The rock, blasted away from Mars by a meteor strike perhaps 15 million years ago, was found in the Allan Hills region of Antarctica in 1984. It is thought to have fallen to Earth some 13,000 years ago. Source: NASA.

clusive, but "considered collectively, we conclude that they are evidence for primitive life on early Mars."

The scientific jury is still out but has been skeptical. Criticisms include that the purported fossilized bacteria are smaller than typical terrestrial bacteria, perhaps too small to hold the chemical equipment needed for reproduction, and that the mineral and chemical deposits could be due to non-biological causes or contamination.

Missions to the surface of Mars may settle the question of Martian life in the not-too-distant future. Martians on the hoof are unlikely to roam the arid and almost airless surface, but probes may look beneath the surface for evidence that water and maybe life from better times retreated underground. Finding fossilized remains of microbes seems more likely than finding live ones—or big critters dead or alive—but the evidence for life would have to be incontrovertible.

If we find clear evidence that life ever existed on Mars, it would be a dramatic discovery and strengthen the rationale for searching for life in other places. Life arising twice in our solar system would suggest that it might circle many suns, although not quite prove it. Knowing that impacts have carried rocks from Mars to Earth, and possibly visa versa, we might still wonder if life arose on just one and spread to the other.

THE GOOD EARTH

The Earth avoided the deep freezer of Mars and the oven of Venus, and not entirely by a lucky distance from the Sun and lucky greenhouse gases in the air. An epic interplay of forces—perhaps including life itself—has kept temperatures in a narrow range here for billions of years.

Without an atmosphere, sunlight would not warm the Earth enough to keep it from freezing. Greenhouse gases in the atmosphere such as water vapor, carbon dioxide, and methane catch infrared light radiated from the sun-warmed surface that would otherwise be lost into space.

A vast cycle of evaporation and condensation moves water and heat around the world, moderating temperatures and bringing water to life on land. Solar heat fuels the engine, turning water into vapor, absorbing heat and forming clouds, and creating temperature and pressure differences in the atmosphere that move the clouds around the globe. Vapor condenses as rain releasing heat and most eventually runs back to the sea, completing its circuit through the water cycle. Clouds have a role in temperature control—when the Earth gets warmer, increasing cloud cover reflects more sunlight back into space, reducing the temperature. When the Earth gets cooler, decreasing cloud cover lets more sunlight reach the surface and warm it.

A much slower carbon dioxide cycle controls the greenhouse effect. Rain and other mechanisms strip the gas from the air, and much of it winds up dissolved back in the oceans, eventually ending up as carbonate rocks like limestone on the sea floor. The process could eventually wash the air of the gas, but over geologic time the rock sinks back into the mantle along with the plates of the Earth's crust and volcanoes eventually recycle some of the gas, spewing enormous amounts of it back into the air. This titanic cycle seems to keep global temperatures from tipping too far off the tightrope range of liquid water.

The Earth appears to have had an extreme ice age three-quarters of a billion years ago, just before the emergence of complex life, with oceans largely and perhaps entirely covered by ice, and land arid and lifeless where it was not covered by glaciers. That's a recipe for falling into a deep freeze like Mars, where a planet's insulating water vapor blanket falls as snow and ice, reflecting more sunlight and making things ever colder. Volcanoes may have saved the day by pumping carbon dioxide back into the atmosphere, creating a super-greenhouse effect that melted ice and allowed the cycle of evaporation and precipitation to begin again.

Life appears to play some role in regulating global temperature, although how important a role is not yet known. The idea is that when temperatures rise, plant life proliferates and converts more carbon dioxide into tissue, and marine organisms thrive and incorporate more carbon into shell. With biology soaking up that greenhouse gas, the Earth radiates more heat into space and cools. When temperatures fall, plant populations decline and fewer organisms chew up carbon dioxide, and more of it builds up in the air, warming the planet. This is a weak version of the so-called Gaia hypothesis advanced by James Lovelock in 1979, which in its stronger controversial form asserts that the entire Earth can be viewed as an organism, substantially regulating the environment and keeping it good for life.

Human activity appears to have raised global temperatures somewhat. Burning hydrocarbon fuels releases carbon dioxide, and it's widely but not universally thought that we have raised global average temperatures by roughly one degree. If intelligent activity can affect the temperature of a planet, then there is reason to think that smart guys might be able to keep planets habitable over very long times—preventing ice ages, for example, or excessive global warming.

HABITABLE MOONS

Planets in habitable zones around stars might not be the only places hospitable to life. Jupiter and Saturn have moons, some three-quarters the size of Mars. Saturn's Titan is 5,150 km, and Jupiter's Ganymede, Callisto, Io, and Europa are 5,262, 4,800, 3,630, and 3,138 km, respectively. These moons are rocky like terrestrial planets, unlike their gas giant parents.

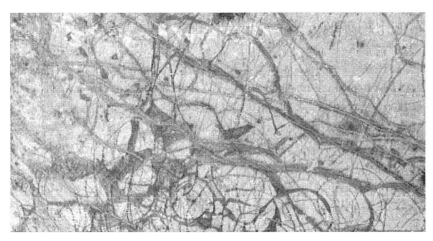

Figure 8.6. Surface of Europa showing a crust of ice, in an image from the *Galileo* spacecraft. Fractures in the ice appear to have been filled in by water or slush. The area in this view is about 360 by 770 kilometers, with features as small as about 1.6 kilometers visible. Courtesy NASA/JPL-Caltech.

In our solar system, the giant planets roam far from the warm zone near the Sun and are extremely cold; Jupiter, for example, is more than a hundred degrees below the freezing point of water. But some of their moons are warmed by tidal heating, with the giant planet's gravity raising tides in the very rock, much like our Moon raises tides on the Earth, and the constant kneading generates heat by friction. Jupiter's Io is so hot inside that it has active volcanoes.

Europa seems to be warm enough to have an ocean covering the entire moon beneath a shell of ice. The best evidence for this is an absence of craters, shown in Figure 8.6. While other moons are pockmarked, Europa is not, presumably because liquid or slush from below fills in craters after impacts. It is conceivable that life exists in water beneath the ice, although very little sunlight would penetrate the kilometers-thick shell to provide energy. Volcanic vents on the ocean floor like the black smokers on Earth might provide an energy source as well as the necessary chemicals.

The possibility that moons of giant planets might provide an alternative kind of habitable zone is intriguing because we know that giant planets exist around other stars, and if they have moons, then there could be more places where life might arise.

SUMMARY

Conditions ripe for life probably exist on some planets orbiting some other stars and possibly billions of them. But nobody has proven that habitable extrasolar planets exist; the idea that some rocky planets with water would fall in habitable zones currently rests on theory, simulation, and statistics. Simulated planets are not as good as real ones (although the real estate is cheaper), so habitable planets are not bedrock certain. Planet searches are detecting Earth-sized planets and may soon measure their temperatures and identify gases in their atmospheres—possibly revealing the presence of water and perhaps even life. Likewise, missions to Mars or moons of Jupiter or Saturn may find signs of life past or even present.

The riot of life we see on Earth gives us reason to think that something similar might break out elsewhere if conditions permit. Its pervasiveness and variety is astonishing: millions of different species, from microscopic to whale-scale; dwelling deep underwater, underground, on land, and airborne; thriving in environments spanning most of the globe; surviving repeated cataclysms over billions of years. Few of us appreciate the variety of terrestrial life until we see pictures of creatures from deep in the ocean, or micrographs of dust mites, bacteria, viruses, and other kingdoms of life just outside of our own daily affairs.

If life does exist elsewhere, the big question is, "Is any of it *smart?*"

Chapter 9

Intelligence and Technology

Searching for signals can only find critters that build gizmos like radios or lasers—intelligent life using technology. That happened once on the Earth, and there is no obvious reason why it could not happen elsewhere. How often it might happen and how long it might take is unknown, but it took roughly four billion years for us to go from algae to atom-smasher—most of the lifetime of our solar system and one-third of the age of the universe.

Evolution

A great deal of evidence supports the idea that we and all other terrestrial life today evolved from earlier, simpler species—ultimately, stuff like algae. In 1858 and 1859, Charles Darwin and Alfred Russell Wallace published evidence that species evolve, with a theory explaining how and why. Until then, species had been regarded as "immutable productions... separately created" as Darwin put it the Preface to his *Origin of Species*. The idea of evolution was not entirely new—the naturalist Jean-Baptiste Lamarck had claimed in

1801 that species, including humans, are descended from other species, and others (including Darwin's grandpa Erasmus) had advanced the idea even earlier. Other people had also suggested mechanisms to explain how evolution worked, although not in much detail.

Darwin presented many volumes of evidence for the mutability of species, and a persuasive theory explaining how it worked—igniting controversy straightaway that still smolders today. At a crowded British Association meeting in June of 1860, Bishop Samuel Wilberforce reportedly asked Darwin-supporter Thomas Huxley "is it through your grandfather or your grandmother that you claim descent from a monkey?" Ouch!

Evolution is driven by genetic variation, which yields random new inheritable traits, and by natural selection, which is the increased chance of a creature reproducing if it has a trait conferring some advantage in the struggle for survival.

Genetic variation is a change in a creature's genetic makeup. It happens through the combination of parental chromosomes and through mutation due to background radiation or some chemicals, as well as other causes. An altered gene passed on to offspring may be crippling or fatal if it interferes with an essential function, but sometimes the gene has a beneficial effect—making the creature better-adapted to its environment, for example.

Natural selection determines whether traits get passed on to future generations. The phrase can be misleading; nature does not somehow recognize and select "good" traits. Rather, individuals that survive long enough to reproduce simply pass on their traits, while those that don't live long enough do not—or the less fit add fewer offspring to the future population. After many generations, much or all of a population may have an originally rare helpful trait, while harmful traits may be selected out.

What's good depends in large part on the current environment; natural selection produces different results under different conditions. In one celebrated example, the peppered moth *Biston betularia* was common in pre-industrial England and liked to perch in trees, most of them light-colored and with a texture that blended in with bark and lichens. When coal-fired industrialization spread soot across the landscape, the light-colored moths became easy for predators to see against darkened backgrounds, and their number plunged while a darker-hued variety became common. Natural selection had reduced the number of moths with genes for light coloration, leaving those with genes for a darker hue to reproduce more often and become more numerous because they were better adapted to the changed environment. As cleaner fuels replaced coal and things became less dingy, the moth population became predominantly light again because darker ones were then easier for predators to see and dine on.

Evolution has no grand design as far as we know, although the results sometimes seem like it does. One example is bacteria that are resistant to certain antibiotics, almost as though they are scheming to foil our medical advances. Many antibiotics work by interfering with some aspect of bacterial chemistry, but an occasional bacterium is not susceptible because its genes code a slightly different chemistry. Those resistant bugs are the ones left to reproduce after an antibiotic kills off the others, resulting in an adapted population that's resistant to our medical miracles.

INTELLIGENCE

Intelligence is often defined as understanding—including reasoning, abstract problem solving, self-awareness, language use, and so on. Other species seem to have degrees of intelligence, but we *Homo sapiens* seem to have a lot more—making things like aircraft, books, computers, dragsters, elements, and the rest of the alphabet of creations—unlike any other species on Earth.

Many species do very nicely with little or no smarts; there are lots of other ways to get by and multiply. The zillion-seed approach works for many plants, insects, and fish that aren't big on brains; only a small fraction of offspring survive long enough to reproduce, but enough to keep the show going. If life was a contest to produce the largest number of individuals or species, intelligence would not win; microorganisms like bacteria are the most successful in sheer number of individuals and total mass, and insects are the most successful in number of species (millions).

Why, then, would intelligence arise? It seems to have some survival value, with smarter creatures better able to avoid being "selected out" and adapting faster to changing conditions. In humans, a great capacity for learning and reasoning helps individuals discover new survival techniques and teach those tricks to others, including their young, neighbors, and eventually large populations as new ideas diffuse over wide areas.

Evolution has produced increasingly complex critters over time. There's no reason to think it must, but that's one result of the last half-billion years. More complex organisms do not necessarily replace simpler ones; they typically coexist with them and sometimes depend on them. Animals, for example, need simpler plants to produce oxygen to breath, and often for food.

Animals have developed increasingly large and complex brains, and intelligence seems to increase with brain size—or more precisely, with the amount of "extra" brain over some basic amount needed to run a body of a given size, as illustrated in Figure 9.1. Modern human brains average about 1.5 kilograms (3 pounds); some elephant brains are 5 kg and some whales are up to 8 kg, but their bodies are much bigger and apparently take more brain to operate.

Brains contain neuron cells, each connected to many others. Nematode worm brains have 302; fruit flies have around 100,000 or ~10^5. Human brains

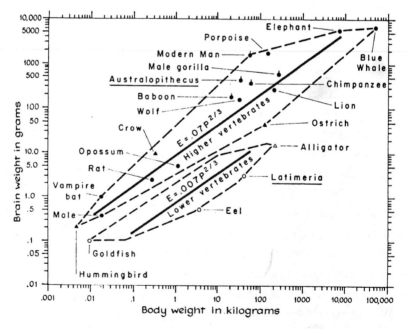

Figure 9.1. Brain and body mass of selected species (dots), with fitted lines for animals of various body mass (based on more cases than shown, and *not* using the regression method). The vertical distance of a species above or below the line corresponds roughly to common impressions of higher or lower intelligence. Courtesy of Harry Jerison.

have $\sim 10^{12}$ neurons, each connected to $\sim 10^3$ others, for a total of $\sim 10^{15}$ connections. Learning seems to make new connections and reinforce old ones, creating memories and associations; memory calls back the stuff encoded in the network. Thought is thought (!) to emerge from the vast number of interconnections, with various functions localized in specialized parts of the brain. A big brain or something like it in complexity seems a prerequisite for intelligence like ours.

Hominids—bipedal primates—appeared roughly five million years ago in the fossil record. *Australopithecus,* with skulls big enough for 400-gram brains (chimp-sized), appeared some four million years ago and extended over roughly two million years, perhaps using crude stone tools. *Homo habilis,* with 800-gram brains, appeared roughly 2.5 million years ago, along with signs of stone tool use and perhaps speech. *Homo erectus,* with 1,000-gram, brains lived from about two million until 250,000 years ago, used fire, and lived in caves. An archaic *Homo sapiens* with 1,200-gram brains showed up toward the end of that time, and modern *Homo sapiens* appeared roughly 100,000 years ago, with 1,500-gram brains (*sapiens* comes from Latin for discerning or wise). Human brain size has not changed much since then, although the most dramatic signs of intelligence are much more recent— writing, for example, is only known to go back about 6,000 years.

In some respects, humans are not very different from other species. The human genome has between 20,000 and 25,000 genes, and appears to be 94% to 98% the same as chimpanzees.

Crudely speaking, men evolved from something much like monkeys in a few million years, but monkeys and everything like them in size and complexity took five hundred million years to evolve from single cells.

Is human intelligence a freak occurrence? There is at least one reason to think that human-like intelligence is not unique. A critter called Neanderthal (*Homo neanderthalensis* or *Homo sapiens neanderthalensis*) roamed Europe and parts of Asia for more than 100,000 years before modern *Homo sapiens* arrived from Africa—thicker-boned than modern humans, and with a more ape-like skull, but with room for brains the size of ours and larger. Fossil evidence shows that they walked upright, made stone tools and spears, used fire, hunted large animals, and buried their dead—eerily almost-human. Comparisons of their DNA with humans suggests that some interbreeding with modern humans occurred. They disappeared roughly 30,000 years ago, perhaps due to absorption through interbreeding, competition with modern humans, changing climate, or other factors. We don't know if they might have eventually invented things like radios or day-time television, but they seem to have been on a path similar to ours.

CHANCE

Events like asteroid or comet strikes, volcanic eruptions, and other misfortunes put zigs and zags into the course of evolution. The fossil record shows repeated mass extinctions where many species suddenly disappear.

In the best-known example, the dinosaurs disappeared 65 million years ago, along with more than half of the other species on the planet, at the geological Cretaceous-Tertiary or K-T boundary ("K" is a geologist's abbreviation for Cretaceous, which comes from Latin *creta* for chalk, which in Greek is *kreta*, which makes perfect sense). Luis and Walter Alvarez and others suggested in 1980 that large amounts of the rare element indium found in rock at that boundary might have come from a big asteroid impact, since asteroids have a high concentration of the stuff, but they were met with skepticism. Ten years later, the 180-km Chicxulub crater beneath the Gulf of Mexico and the Yucatán peninsula was discovered and dated to the same age, which has convinced many that the mass extinction was due to an asteroid impact.

The impact of an object 10 kilometers across could bring global disaster in many ways. An immediate effect would be intense heat vaporizing huge volumes of ocean and setting nearby land areas on fire, followed by massive tidal waves. Smoke and clouds blocking sunshine could cause a global winter that might last years, with reduced sunlight starving plants and in turn starving animals. Lower temperatures might be especially tough on cold-blooded animals, and might favor warm-blooded creatures thought to have been overshadowed by reptiles until then.

We might not be here if not for dinosaurs' bad luck.

Does human existence turn on that single event? The dinosaurs' downfall was not a unique event; mass extinctions have occurred repeatedly. The Permian-Triassic "Great Dying" 250 million years back, for example, snuffed

90% of all species, and other worlds might suffer similar extinctions due to impacts, volcanism, or other catastrophes. We are the product of one chain of events that would not be duplicated exactly anywhere else, but many planets might have similar histories with catastrophe occasionally stirring the pot.

If we were not here, something else smart might be, perhaps descended from dinosaurs or other lines; and if not yet, then maybe after another roll of the dice.

Mass extinctions underscore the sobering facts that life is fragile and that few if any species survive forever; fewer than one in a hundred species that ever lived is thought to be around today. Intelligence and technology might change that—heading off catastrophic impacts and inventing solutions to other perils so that wise guys might escape outrageous fortune.

TIME

Human intelligence appeared 4.5 billion years after the Sun and Earth formed, which gives us a clue as to how long intelligence might take to emerge elsewhere. That's nice to know, because stars and their planets have an enormous variety of ages ranging from just a few million years to around thirteen billion. Very young stars have probably not been around long enough to harbor home-grown intelligent life, but a fairly large fraction of stars are as old as the Sun and many are billions of years older—enough time for intelligence like ours to evolve.

The universe is thought to have begun with a Big Bang about fourteen billion years ago, when pure energy began expanding from an almost infinitely hot and dense point. Cooling as it expanded, elementary particles like quarks formed, and then protons and neutrons, combining after some 300,000 years into light atoms like hydrogen along with some helium and lithium, but none of the heavier elements. After less than a billion years, vast clouds of this stuff were drawn together by gravity to form the first galaxies and stars.

This scientific story of how things came to be is backed by evidence that has convinced most scientists. One supporting fact is that the proportions of hydrogen and helium observed in the universe very nearly match what theory predicts a big bang would cook up—it's 75:25 by mass, or 12:1 by atoms.

Another supporting fact is that most galaxies are flying apart from each other, known because their light is Doppler shifted toward the red end of the spectrum. Perhaps the strongest support is that a glow can be detected in all directions at microwave and infrared wavelengths, and its temperature is almost exactly the 2.7 degrees Kelvin predicted for a flash of creation after expanding and cooling for billions of years. This cosmic microwave background radiation was first detected by Arno Penzias and Robert Wilson in 1965 with a radio telescope, and it's thought that it could be so uniform across the sky only if everything was once in a very small volume, like a big bang fireball. In the early 1990s, the Cosmic Background Explorer Observatory (COBE) satellite measured the radiation very precisely and found it to match the spectrum expected of a big bang, and to be within one-hundredth of one percent of the same temperature in all directions—the slight variations thought to be the cause of clusters of galaxies forming in the early universe.

The oldest stars might seem the most likely to harbor intelligent life since it would have the longest time to evolve, but they are low in metallicity—astronomers' zany jargon for elements heavier than hydrogen and helium. Physicists are pretty sure that big bangs can't cook up the heavier elements, so the earliest stars probably don't have rocky planets like the Earth, with carbon and other stuff needed in the familiar chemistry of life. While the oldest stars would allow the longest time for intelligence to evolve, they don't seem very likely sites for life to get started in the first place.

A theory of stellar evolution tells us the age of stars and their fates. It's called "evolution" because stars change size, temperature, and composition as they age, looking very different at various stages in their lives. We know the intimate secrets of these distant giants because emission and absorption lines in their starlight spectra reveal the temperature and composition of their outer layers. Temperature tells the star's mass because fusion fires burn hotter as more mass is piled on, increasing the pressure and temperature at the core where fusion occurs.

Life expectancy is mostly a matter of mass.

Stars much less massive than the Sun burn cooler and slower. The lightest stars, less than one-tenth of the Sun's mass, are expected to shine for hundreds or thousands of billions of years, which is much longer than the current age of the universe. Less massive bodies don't shine at all because they don't compress hydrogen enough for fusion. Stars down to about one-half the Sun's mass are thought likely to have habitable planets.

Our Sun is near the middle of a ten-billion year lifetime typical of stars its size. The bad news is that it's getting hotter, and in a billion years may make the Earth too toasty for life to survive. Stars like ours eventually become red giants as they exhaust their hydrogen fuel, expanding roughly a hundred times in size. The Sun is roughly a million miles across today (1.4 million kilometers), and roughly a hundred million miles away—only one hundred solar diameters distant. Expanding a hundred times in its red giant phase over several billion years, the incandescent surface is would come regrettably near the Earth. Happily, that's millions of human generations in the future, and clever creatures might cope by moving further out in the solar system, or by moving to other stars, or by strategies we can't imagine.

It's sobering to realize that we came on stage halfway through Act Four in a drama that may have only five billion-year acts!

Stars much more massive than the Sun burn hotter and live much shorter lives. Twice as massive, and they last only about a billion years—perhaps long enough for life to evolve, but probably not complex creatures if our history is a good guide. Eight times or more massive than the Sun, and stars live only a few tens of millions of years.

The short life of massive stars is where heavier elements come from—the stuff big bangs can't brew. When hydrogen fuel runs too low to counter the crush of gravity, a star contracts, pressure and temperature rise, and helium begins burning. Helium nuclei fuse to make nuclei of heavier elements such as carbon and oxygen, plus energy, and those elements are later squeezed together to feed the fusion fire, making even heavier nuclei up to iron. Iron, however, is the end of the road—iron fusion won't heat a star enough to counter gravity. The star's core collapses and temperatures soar in a titanic supernova explosion synthesizing all the elements up to uranium and beyond, much of which blows off into space to form a cloud-like nebula.

Eventually, some of the material is drawn together to form new stars enriched with heavier elements; the Sun is known to be a second or later generation star because it contains heavy elements cooked up in earlier supernovae. We, everything around us, and most stars in the galaxy are made of stuff forged in earlier generations of stars.

The most promising places for finding intelligent life seem to be stars between half and twice as massive as the Sun, and already several billion years old. Such stars have been forming for more than ten billion years, over twice the age of our Sun, so there are many where life may have had a lot longer to get smart than we have.

Naked apes with three-pound brains figured out what makes stars shine and proved it with hydrogen bombs, and those sometimes comical creatures also puzzled out how the universe itself was born and the fate of its stars. We would be awfully conceited to think others elsewhere might not be as clever as we are (although they probably would not be as good-looking).

TECHNOLOGY AND COMMUNICATION

Searching for signals from other worlds means looking for wise guys using technology, not merely intelligent creatures. We don't know how often the two go together, and it's probably not inevitable, but in our own case technology seems a natural outgrowth of intelligence, one that grew fabulously fast once it sprouted.

All human cultures have technology—mechanical arts—making useful objects, especially tools, by applying knowledge of how things work. Its roots are in basic needs for food, shelter, warmth, and the like; the knowledge of how to hunt and farm, build dwellings, make fire, and so on has helped humans spread across the globe. Technology apparently has survival value: the tremendous increase in human population and life expectancy in recent centuries is largely due to technology and its scholarly sibling, science.

Technology might not be an inevitable outcome of intelligence; dolphins and some primates seem fairly smart but lack it. In the case of dolphins, developing technology under water might not be feasible, and the motivation might not be as strong as for land-dwellers; ocean worlds might not breed radio builders. On land we see some simple tool use among some other species, such as using sticks to get at food, but none make tools like we do or use them so constantly.

Language and culture are among the most striking differences between humans and other species with some intelligence.

Language provides symbols to represent many things including knowledge and communicate it to others, and while some other animals are capable of simple communication like alerting others to danger, we can communicate very complex things. Human adults know between 5,000 and 50,000 words (experts can't agree if words like *drive*, *driver*, and *drives* should be counted as separate), and such complex spoken language requires fairly complicated vocal organs along with a certain threshold of brain size and specialization which seems unique to us on Earth.

Culture includes (among other things) a body of knowledge greater than a single individual can acquire, remember, use, or pass on, and it's what allows knowledge to span many individuals, places, and generations. Technology is one part of culture and one part of technology is the know-how that allows us to record language in writing and other media, make copies of it, share it with others, and store it for future use.

It's easy to imagine a civilization that never progresses past wooden, stone, or simple metal tools, and until recently that was the human condition. But there seems to be an inexorable logic to technological development, which led quite quickly from wooden cartwheels to steel wheels rolling on rails, from runners carrying news for days to radios flashing it around the world in less than a second, and from sailing ships to spaceships. It's hard to imagine smart guys not constantly seeing better ways of doing things, which is how we arrived at our current "high" level of technology.

Contrary to popular impressions, radio is not a particularly high technology; the basic ideas and devices appeared in the era of horse-drawn wagons, steam locomotives, and chamber pots. James Clerk Maxwell predicted electromagnetic waves in 1864 and published a complete theory explaining them in 1873. In the late 1880s Heinrich Hertz discovered that when a spark jumped a gap in a simple electrical circuit, a spark appeared in a tiny gap between the ends of a separate loop of wire some distance away. Electromagnetic waves carried energy through the space between them, radiating from the transmitting circuit as electrons surged back and forth during the first spark, and when the waves rippled across the receiving loop their field generated voltages in it making the second spark. No fancy vacuum tubes, transistors, or integrated circuit chips—just loops of wire and some electricity.

Electricity is not very high-tech either; it had been known for over a century before radio was invented and was brewed in jars of chemicals.

Radio waves were sent over increasingly long distances during the 1890s, and by early 1902 Guglielmo Marconi sent Morse code across the Atlantic Ocean. Commercial broadcasting began in 1920. One hundred years after spanning the Atlantic, we now have two-way radio contact with spacecraft at the edge of the solar system, and can detect the faint radio hiss from objects billions of light years away.

It is absolutely certain that radio and other kinds of electromagnetic signals like light can span the enormous distance between stars.

One thing that makes electromagnetic waves so attractive is that they keep going *forever*. They do weaken with distance, because they spread out over an ever-larger area as they move away from their source, rather like an inflating balloon getting thinner as it expands. At twice the distance from the source, they are only one-quarter as strong over a given area (say a square meter) and at ten times the distance, one-hundredth as strong, following the inverse square law $1/R^2$ where R is range. Pleasantly, they travel at the speed of light—nothing's faster. Radio waves pass through space freely, although some wavelengths are absorbed by some molecules in interstellar space and by stuff like water vapor in atmospheres; light passes through space freely too although it can be dimmed or extinguished by interstellar dust and thick atmospheres or clouds. Electromagnetic waves seem to be the best way nature offers for signaling over long distances, which is a good reason to expect that other technological critters would use them as we do.

TECHNOLOGICAL LIFETIME

How long do technological civilizations last?

If they only last a few hundred or thousand years, then we might be the only one in the entire galaxy at the moment. Even with a million-year lifetime, thousands could have come and gone over the last billion years, with few existing at the same time—or now.

The average lifetime of terrestrial species is thought to be roughly ten million years, but that statistic might not be relevant to wise guys like us. Many human civilizations have come and gone in less than a thousand years; we know of none that have lasted five thousand although China's written history goes back 3,500. Viewing the various cultures and nations and empires over the ages as parts of a single human Civilization, it goes back at least 5,000 years to the earliest writing by Sumerians, and more than 10,000 years to early agriculture. If we view the earliest actors in the *Homo* line as the pioneers of technology for using stone tools and fire (but not agriculture or cities), then our Civilization extends back roughly a million years, although it's only been technological in the sense of radio for one hundred.

It's customary to note that our civilization could be destroyed by some of its creations such as nuclear weapons, but unless our entire species was wiped out, another human civilization might very well rise. Our species might last through the rise and fall of many civilizations over millions of years, although not all of them might develop stuff like radio.

Over thousands of years, civilizations may be cut down by depletion of resources (fossil fuels, arable land, accessible ores, and so on), changes in climate (ice ages, global warming, drought), and many other physical perils. Cultures change over time too, and some might abandon technology in favor of mysticism, hedonism, or fashions we can't imagine. On a million-year time scale, smart critters might evolve into something quite different or transform themselves through genetic engineering. Over hundreds of million years, global catastrophes like asteroid or comet strikes loom more likely; over billions of years stars change and some die. Civilizations would need to dodge a lot of bullets to live indefinitely.

Some might, however, last a very long time. One reason to think so is statistical—if many have been born and their lifetimes vary randomly, then some should last a lot longer than average. Another reason is that intelligence and technology might be used to deflect killer comets, avoid catastrophic wars, and so on. Civilizations might not remain tied to the fate of a single planet or star; our current exploration of the solar system may be the first steps in human migration away from our home world, and clever creatures elsewhere might follow similar paths.

It is possible that something intelligent but not alive might remain after a smart species disappeared. Given enough processing power, memory, and interconnections, it's conceivable that artificial intelligence could be created in computer-like forms, possibly mimicking biological intelligence and perhaps someday superseding it. Thinking machines have yet to be built, but specialized hardware and software can now match or out-perform people at some complex tasks like chess and even the game Jeopardy! The fast-growing power of microelectronic circuits makes artificial versions of intelligence seem possible; with enough time, money, and graduate student labor, that possibility might be realized here or elsewhere.

It is conceivable that brains—memory, thought, personality; the whole mess—could be copied into artificial forms like computers. Roboticist Hans Morovec called the result Mind Children and estimated that the technology might be ready before the middle of the twenty-first century. I'd call my copy or copies "Bob in the box" and presumably we would argue about which was the real me.

The notion of extraterrestrial intelligence and civilizations should, then, include the possibility of inanimate but intelligent entities. Whether smart machines would have motivation for communication is anybody's guess; perhaps they would get some form of satisfaction from establishing data links with distant network nodes on other worlds.

DRAKE'S EQUATION

Astronomer Frank Drake cobbled up an equation for a meeting at Green Bank, West Virginia, in 1961 to estimate how many communicative civilizations might exist in our galaxy at any one time, shown below with the terms given more memorable names:

$$N = R^* \, f_{planets} \, n_{habitable} \, f_{life} \, f_{intelligence} \, f_{signals} \, L$$

where N is the number of potentially detectable civilizations, R^* is the annual rate of formation of suitably long-lived stars, $f_{planets}$ is the fraction of those stars that have planets, $n_{habitable}$ is the number of potentially habitable planets or moons per star, f_{life} is the fraction of those habitable places where life appears, $f_{intelligence}$ is the fraction where intelligence evolves, $f_{signals}$ is the fraction using electromagnetic signals that we could detect, and L is the number

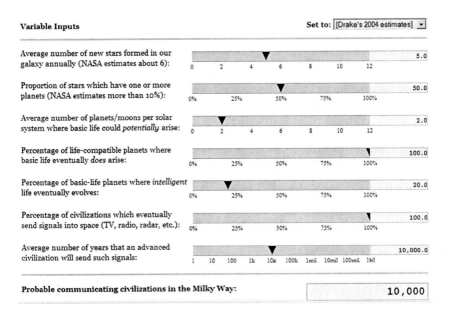

Figure 9.2. One set of assumptions yielding 10,000 potentially communicative civilizations in our galaxy, from a Drake equation calculator by Alan Bellows.

of years they remain detectable—presumably as long as a technological civilization lasts. A link to an interactive calculator is given at the back of this book, with sample results shown in Figure 9.2.

We know enough to estimate the first three terms. Suitable stars are thought to form at the rate of between one and ten per year in our galaxy. Planets circle well over ten percent of the stars studied so far, and theories of stellar and planetary formation suggest the fraction may be closer to 1. The number of habitable environments per planetary system may be near 1, and it could go higher—we've got the Earth, plus the near-miss Mars and Venus, plus several long-shot contenders in some moons of Jupiter and Saturn. The first three terms yield numbers between 1 x 0.1 x 1 = 0.1 and 10 x 1 x 1 = 10; call it roughly 1.

The next three terms are currently unknown. If optimists are right that life, intelligence, and communication technology are all pretty likely on habitable planets, then those three terms might all be close to 1 and the formula would become roughly N=L, so the number of communicative civilizations in the galaxy is roughly equal to their average lifetime. If that lifetime is millions of years, then there could be millions of them. If the lifetime is only a hundred years (because some catastrophe always deep-sixes wise guys), then there might be one hundred scattered throughout the galaxy—but none might happen to be in our 1,000-light-year neighborhood which includes very roughly 1/1,000th of the stars.

Pessimistic estimates of the likelihood of life, intelligence, and technology paint a bleaker picture. If the odds are one-in-a-hundred for each of those terms, then this part of the equation is 10^{-2} x 10^{-2} x 10^{-2} = 10^{-6}, and the first three terms don't matter much because this part is so small. If the typical lifetime is a million years, we get 10^{-6} x 10^6 = 1 communicative civilization in our galaxy, and more pessimistic estimates yield mere probabilities that we share the galaxy with anyone.

Drake's estimate in 2008 was 10,000 (5 x 0.5 x 2 x 1 x 0.2 x 1 x 10,000). My own guess is that fairly simple life might be widespread, and that intelligence and technology are rare but might last a very long time in some cases. It's hard to plug such fuzzy notions into the equation, but if lifetime is very large (say, a billion years), then even pessimistic estimates for other terms yield thousands of potential communicators.

The formula does not include some terms that might be important, because we don't know all of the factors that should be included or what values to assign them. For one example, it does not consider the effect of growth—the possibility that evan a few civilizations might colonize many stars over time. We need to learn more to fill in the unknowns, such as the likelihood of life, and may find it on the barely-explored planets and moons of our solar system, or in the spectra of extrasolar planet atmospheres. It's unlikely, though, that we will discover much about the likelihood of intelligence, technology, or life expectancy without searching for electromagnetic signs that somebody is out there.

SUMMARY

Intelligence might exist on some other worlds if life has developed and complex creatures have evolved. How often intelligence anything like ours might emerge is unknown—it's probably not inevitable, and it might be rare, but we know that it is possible because here we are thinking about it.

Technology may not always develop, but human intelligence produced technology rather quickly. Basic technology such as wood, stone, or metal-working may not always lead to radio, but it happened here in a matter of millennia—the blink of an eye on the cosmic time scale. Radio or similar signaling can make life detectable across interstellar distances, especially if it wants to be found. If broadcasters are common, then we might find little green guys with our current modest level of effort; if they are rare, then we might need to look at a great many stars or eavesdrop with much more effort.

CHAPTER 10

INTERSTELLAR COMMUNICATION

SPANNING THE GALAXY

A remarkable fact first noted by Carl Sagan and Frank Drake is that we could detect a radio signal *from anywhere in our galaxy* of several hundred billion stars by listening with our 1,000-foot antenna at Arecibo, Puerto Rico (shown in Figure 10.1), provided a similar antenna was pointed in our direction radiating no more power than we could easily afford.

One hundred years after we first spanned the globe with radio, we could span the entire galaxy!

There are a few practical problems.

One problem is that the two antennas need to be pointing at each other. Arecibo's gain, or amplification in the direction it's pointed, is a factor of roughly ten million, but only over the ten-millionth of the sky directly in front of it, and the distant transmitting twin would need to illuminate us in a similarly tiny spot of sky. It's unlikely that two such big antennas would

Figure 10.1. The 1,000-foot (305-meter) radio telescope at Arecibo, Puerto Rico. The spherical reflec-tor focusses radio waves onto antennas suspended 450 feet above. Since the reflector can't move, the suspended antennas are moved to track objects within some 20 degrees of the zenith. Cour-tesy of the NAIC - Arecibo Observatory, a facility of the NSF.

ever point at each other by chance, but if we knew each other's location and frequency, it could work.

Another problem is that a lot of power would be needed to make a sig-nal detectable across the galaxy—roughly the output of a big power plant, or 1,000 megawatts (a megawatt is a million watts, abbreviated MW)—and sending information at a reasonable rate might take 10 or 100 times more power depending on details. Things flying over such a transmitting dish might cook! A different kind of antenna system called a phased array would probably be used for transmitting, which would solve that problem by spread-ing the power over many small antennas scattered over a large area.

Signaling across our 100,000-light-year galaxy is an extreme example, but it illustrates the immense reach possible with our current technology. ET might reside much closer to us because millions of stars are within just 1,000 light years of the Earth, and thousands are within 100 light years. We could reach many of them with the one-million-watt transmitter at Arecibo, normally used for planetary radar, and shorter distances would require *as little power as some kitchen appliances.*

WHY RADIO?

Electromagnetic waves are attractive for signaling over cosmic distances because they are easy to generate and detect, and because they travel at the speed of light. The radio portion of the spectrum has been thought to be the most promising, but the optical portion has attractions as well. Other ways of signalling are possible, such as sending particles or even rockets, but all have big drawbacks. It is conceivable that new forces or methods will be discovered that make electromagnetic communication obsolete, but few physicists are holding their breath.

One attraction of radio is that most stars are dark at those wavelengths, so signals would not be obscured by a bright background. Optical signals, on the other hand, would need to outshine the home star's glare to be detectable. Another advantage of radio is that space is fairly transparent to it at many frequencies, while light is dimmed by dust in space; the Milky Way would be a much brighter band across the sky without interstellar dust.

There is a radio window through space from about 1 to 60 gigahertz where natural noise would not interfere with signals very much (a gigahertz is 10^9 cycles per second, abbreviated GHz, and frequencies this high are called microwave; cycles-per-second has been dubbed Hertz after Heinrich Hertz, who first demonstrated radio). At lower frequencies, the radio sky brightens with background noise from the galaxy, which contributes to static on radio and television, and at higher frequencies receivers generate more internal noise. This free space microwave window is the same for everyone every-where—a naturally quiet part of the spectrum where signals would not have to compete with much background noise.

Another reason to prefer frequencies in this range is purely practical and perhaps not universal: our microwave radio technology is very good. Massive research and development efforts in radar and radio astronomy have brought microwave receivers close to perfection; the best add very little internally generated noise to the 2.7 K cosmic background. Yet another reason to favor radio is that it penetrates clouds, while ground-based optical communications can be blinded by bad weather.

TERRESTRIAL MICROWAVE WINDOW

Our atmosphere is like a dirty window across some parts of the radio spectrum for observers on the ground. Water vapor, oxygen, and other stuff absorbs energy in some bands, adding background noise that could drown out signals. But there is a part of the spectrum where both cosmic and atmospheric noise is fairly low, called the terrestrial microwave window, covering roughly one to ten GHz and shown in Figure 10.2. If broadcasters seek an audience of oxygen breathers on the surface of wet planets, frequencies in that range seem especially attractive.

Most searches have covered only thin slivers of frequency, often less than one megahertz listening near the 1.42 GHz hydrogen frequency, which is only one ten-thousandth of this window. Hydrogen's radio emission is itself a background noise—it glows bright in the plane of the galaxy and is detectable in all directions. The frequency of some other emission lines in space have also been proposed to guess ET's phone number—one is the hydroxyl

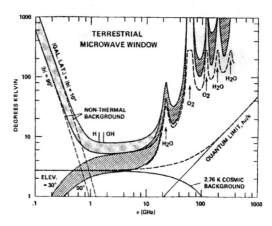

Figure 10.2. The terrestrial microwave window between 1 and 10 GHz. Hydrogen's emission is marked "H", while "OH" marks several close-spaced spectral lines of hydroxyl. Frequency is shown along the horizontal axis in GHz (1,000 MHz) and background noise is shown on the vertical axis in degrees Kelvin. Source: NASA SP-419.

or OH molecule that glows near 1.66 GHz and there have been many others. The spectrum between the H and OH lines is sometimes called the "water hole" because both are components of water; it's a relatively quiet band where some have poetically suggested that different species might meet, just as some Earthly creatures gather at real water holes to drink (it's a stretch). Listening at such magic frequencies has the problem of additional noise, but it's the band around the magic frequency where searchers usually tune in.

The Allen Telescope Array discussed later will be the first to cover the entire window, avoiding guesses about frequency.

SPECTRUM ANALYZERS

It's thought that interstellar radio beacons would have very narrow bandwidths because narrow signals would be obviously artificial, and are fairly easy to detect with multi-channel spectrum analyzers even if they are weak. Unfortunately, as we slice the spectrum into narrower channels, we get many millions of them.

There's a limit to how narrow radio signals can be after traveling enormous distances through space because even a perfectly monochromatic signal at a single frequency gets spread out a bit along the way. Variations in the density of electrons in space cause slight zig-zags in direction, so that some parts of a wave arrive slightly later than others—spread out in time and therefore in frequency. The effect is roughly one-tenth of a Hertz over a thousand light years, first noted by Frank Drake and George Helou, so slicing the spectrum into channels that size is about as fine as it's worth dicing (or narrower if you think ET is closer).

Multi-channel spectrum analyzers make it possible to monitor millions of channels at the same time, vastly speeding up searches. They are based on computer programs that calculate the Fourier transform of the receiver output, which is simply a voltage varying with time, to produce a power spectrum that shows the amount of power at each frequency. Figure 10.3 shows some examples.

Unfortunately, spectrum analyzers occasionally spew out big numbers—brief snap, crackle, and pops that occur when a great many measurements

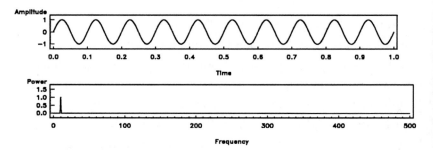

Figure 10.3a. A simulated signal passing through ten cycles during a length of time (top), with its power spectrum (beneath) showing a spike at a frequency of 10 cycles, both detecting the signal and giving its frequency. The frequency scale goes up to 500 cycles because the amplitude of the simulated signal was sampled 1,000 times, and signals at any frequency up to half of the sampling rate could have been detected.

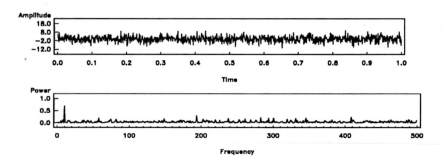

Figure 10.3b. The same simulated signal with three-times-stronger random noise added to almost completely obscure it (top), with its power spectrum (beneath). The 10-Hz signal is still clear in the power spectrum despite added noise.

are made in the presence of noise—so everything below a false alarm threshold decided by statistics is usually ignored. The threshold is sometimes 20 times the noise level for million-channel analyzers running for many hours, although it can be set lower if many measurements are averaged to dampen big spikes. Averaging can be dangerous, though, because a strong signal present only part time could be obscured by being averaged with lots of noise.

Spectrum analyzers are terrific at finding even very weak signals, but the signal must be strong enough to stand above the ever-present background noise, much like a voice needs to be loud enough to be heard above other conversations in a crowded room. To detect a signal across interstellar distances, transmitters need to illuminate receivers with enough power to stand above the background babble of noise.

RANGE

What does it take to signal between stars?

The huge distance between transmitter and receiver is the main problem because the energy falling on a distant receiver antenna decreases with the square of the distance—the same receiver ten times further away receives only one-hundredth of the signal power, which might not be enough to detect. The broadcaster can help compensate for distance by sending more power, or by using a bigger antenna to concentrate the power in a narrower beam, or both. The searcher can help compensate by using a bigger receiver antenna to collect more power or by using a quieter receiver, and to a lesser extent by adjusting the width of channels and integration time (formula are given in the back pages).

Broadcasts between nearby stars require surprisingly little power, as shown in Table 10.1. With Arecibo-scale antennas used for both the broadcast and receiver, one kilowatt (1,000 watts, the power used by a microwave oven or toaster) would be detectable across ten light years, which would reach a few nearby stars, and a hundred kilowatts (comparable to a television station) would be detectable over a hundred light years, which contains thousands of stars. A million kilowatts (comparable to a big power plant) would have a range of 10,000 light years, covering billions of stars. Even a smallish 30-meter receiving antenna could detect that signal over a thousand light years, and a

Table 10.1 Range of a 300-meter Antenna Broadcast for Various Transmitter Power and Receiver Antenna Sizes

Transmitter Power		Receiver Antenna Diameter			Stars in range of 300-meter receiver antenna
		3 meters Home-size	30 meters Medium-size	300 meters Arecibo-size	
Watts	Example	Range (ly)	Range (ly)	Range (ly)	
1,000	microwave oven	—	—	10	few nearest stars
100,000	television station	—	10	100	thousands
10,000,000	big radar	10	100	1,000	millions
1,000,000,000	big power plant	100	1,000	10,000	billions

Frequency 1.4 GHz, 25 K system temperature, 1 Hz channels, 1 sec. integration, signal-to-noise 9, antenna efficiency 0.5

backyard-scale 3-meter antenna would have a hundred-light-year range. We could receive signals from billions of stars, assuming a big enough antenna is aimed at us.

These examples assume a signal only 1 Hz wide, which is nicely matched to our narrowband receivers, but does not deliver information very fast; information would flow at one roughly bit each second which is slower than Morse code. You could receive $E=mc^2$ in less than a minute, but hopefully there would also be a much faster signal with richer content. Assumptions about bandwidth and other details are shown in the fine print beneath each table and are rather conservative. For example, integrating for hours instead of the one second assumed here. along with some other changes, could increase the range by ten times—making one power plant detectable across the galaxy, rather than a the 10,000 light years shown here.

ONE POWER PLANT

We have just seen that it's possible to signal over interstellar distances if broadcasters use a big antenna, but we don't really know what size antennas they might use or how much power is available. Table 10.2 shows the range of broadcasts for a variety of antenna sizes ranging from tiny to much bigger than Arecibo, holding the amount of power constant at 1,000 megawatts or one power plant. That much power costs millions of dollars per day here, which seems pricey, but it's useful for illustration.

If broadcasters use very large antenna systems—say, 3,000 meters in effective diameter, much bigger than Arecibo—then they could be detected by a backyard dish listening a thousand light years away, or anywhere in our galaxy by an Arecibo-scale dish, or in nearby galaxies by searchers using similarly large antennas. Such big antenna systems are certainly possible; NASA's 1971 *Project Cyclops* study estimated an array that size might cost ten billion of dollars, and radio astronomers are currently planning a Square Kilometer Array.

If broadcasters use omnidirectional antennas to broadcast in all directions, a single power plant is not detectable at even nearby stars unless the receiving antenna is bigger than our biggest at Arecibo—and even 3,000 meter receiv-

Table 10.2 Range of a 1,000-Megawatt Broadcast for Various Antenna Sizes					
		Receiver Antenna Diameter			
		3 meters Home-size	30 meters Medium-size	300 meters Arecibo-size	3,000 meters Cyclops-size	
Transmitter						Stars in range of 300-meter receiver antenna
Power (MW)	Antenna Diameter (m)	Range (ly)	Range (ly)	Range (ly)	Range (ly)	
1,000	Omnidirectional	—	—	3	30	none
1,000	300	100	1,000	10,000	100,000	billions
1,000	3,000	1,000	10,000	100,000	1,000,000	100 billions

Frequency 1.4 GHz, 25 K system temperature, 1 Hz channels, 1 sec. integration, signal-to-noise 9, antenna efficiency 0.5

ing antennas would not have a very long range. Broadcasting in all directions all of the time requires far more power than single terrestrial power plants.

OMNIDIRECTIONAL BROADCASTS

Radiating in all directions all of the time, broadcasters need not know the location of searchers, and searchers can assume signals are present all of the time, but it requires an enormous amount of power as illustrated in Table 10.3. Roughly speaking, ten million times more power is needed than in the case of broadcasting with an Arecibo-scale antenna because the ten-million-times gain of the big antenna no longer directs power toward a single star.

The table shows that even if searchers use a big Arecibo-scale antenna, an isotropic broadcast needs 10^{14} watts to reach 1,000 light years, which is comparable to all power use on Earth, and reaching across the galaxy would require 10^{18} watts—a hundred times more power than the Sun shines down on Earth. Even if power was free and so much was available, such broadcasts probably could not be made from the surface of a planet due to waste heat toasting the taxpayers.

Searching for omnidirectional broadcasts means assuming very advanced and highly motivated broadcasters. It's possible that some exist, but it would be nice to avoid making such optimistic assumptions. One alternative is assuming a big antenna pointed our way all the time, which would require far

Table 10.3 Power Required for Omnidirectional Broadcasts for Various Ranges

| Transmitter Power | | Receiver Antenna Diameter | | | Stars in range of 300-meter receiver |
| | | 3 meters Home-size | 30 meters Medium-size | 300 meters Arecibo-size | |
Watts	Example	Range (ly)	Range (ly)	Range (ly)	
10^{10}	10 big power plants	—	1	10	few
10^{12}	1,000 big power plants	1	10	100	thousands
10^{14}	all human power use	10	100	1,000	millions
10^{16}	all Sun's power on Earth	100	1,000	10,000	billions
10^{18}	hundred-millionth of Sun	1,000	10,000	100,000	100 billion

Frequency 1.4 GHz, 25 K system, temperature, 1 Hz channels, 1 sec. integration, signal-to-noise 9, antenna efficiency 0.5

less power and would be just as easy to find as an isotropic broadcast, but it would seem awfully lucky for us to be chosen.

Most searches assume that signals will be present whenever we happen to look—and most searches view any one position for only a matter of minutes, which can find only a continuous omnidirectional broadcast or a broadcast targeting us all of the time. It's possible, however, that omni broadcasts aren't "on" all the time, and we're not a target all the time.

RECOGNIZING SIGNALS

Our local radio spectrum is a babble of man-, woman-, and machine-made chatter, and the sky hisses with natural radio noise. How could we tell interstellar signals from local ones or natural sources?

It's fairly easy to distinguish artificial signals from natural sources. Natural sources are usually noise-like and hiss across the spectrum, while radio signals are usually concentrated in a narrow band of frequency, like a pure tone. Some natural sources such as hydrogen have a single-frequency emission, but it's always spread over a wide range by Doppler shifts because the stuff moves with various velocities over vast volumes of space. The narrow channels used in most SETI receivers are fairly insensitive to natural sources because not much noise falls in any one channel.

Distinguishing man-made transmissions from little-green-man-made ones is more challenging, but there are several ways to tell local signals from those beyond the Earth.

One simple test is to point the antenna away from an apparent source in the sky. If the signal remains, then it's probably local interference; if it disappears but returns when the antenna is pointed back toward the source, then it's potentially interesting. A better test is to see if a source moves across the sky at the sidereal rate of 15° per hour from the east to west, like the stars. Sidereal motion is caused by the Earth's rotation, and a telescope tracks stars by moving in the opposite direction at the same rate; a source that stays in the beam while it's tracking the stars is probably among them—no satellites, planets, aircraft, or other local bodies are likely to mimic that motion.

Celestial signals should also display certain Doppler shifts and drifts in frequency. A frequency shift is expected of signals from outside the solar system because of the Earth's orbital motion and the Sun's motion. The shift amounts to a few kHz each day in the water hole and it changes in a predictable way as the Earth makes its way around the Sun during the course of the year, as shown in Figure 10.4.

A drift in frequency should also be seen—a small, continuous change in frequency caused by the Earth's rotation. An antenna on the Earth's surface is rotating at 1,000 miles per hour at the equator and the velocity along the line

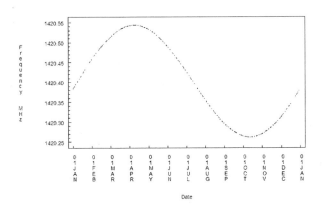

Figure 10.4 The effect of Doppler shifts on a signal from outside the solar system with a fixed frequency of 1420.356 MHz. The shifted frequency (vertical scale) is calculated for a receiver in Chicago, Illinois, pointed at declination -27°, for the time the Wow position transits on each day of the year (horizontal scale).

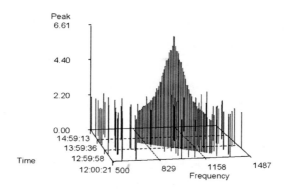

Figure 10.5. A simulated signal drifting in frequency at 6 Hz per minute, typical of Doppler drift caused by the Earth's rotation. The rise and fall of intensity is a separate effect, caused by a constant-intensity source moving through the beam of a transit antenna. Random spikes are noise from real observations.

of sight toward a celestial object constantly changes, causing its frequency to slowly change like a sine wave over 24 hours (because the rotation is in a circle). The drift rate is small, typically 6 Hz/minute at the 21-cm wavelength; a signal with just the right drift rate for its direction would look very much like it was from beyond the Earth. Figure 10.5 shows an example of what such a signal might look like.

CONFIRMING INTERSTELLAR SIGNALS

To be certain that a signal came from the stars, we would want detections by several different observers at different sites using different instruments, which would reduce the chances of being fooled by interference, bugs in software or hardware, or tricks. A signal passing those tests and showing the expected Doppler effects would be strong evidence of ET.

Unfortunately, most radio searches can't tell if an exciting signal comes from a star because most antennas see many stars at once. But telescopes like the Very Large Array could be used to make high-resolution radio images showing the exact coordinates of a signal, and if that position coincided with a star it would be convincing that ET was on the line. Once a signal was announced, lots of scientists would set out to verify and study it; a mistaken claim would not survive long.

An initial detection would probably not yield any message because the spectrum analyzers used for searching typically average signals for about a second, which would lose information arriving faster than about one bit ("beep") per second. That's slower than someone sending Morse code with a telegraph key, and messages might come much faster. To get the message after detecting a signal's presence, we might need specialized receivers or bigger antennas or both.

COMMUNICATING

An initial signal might describe the receiver system used by our prospective pals if two-way communication was desired. They might direct us to a much higher radio frequency or to the optical realm for a faster and more efficient link. Knowing each other's location in space and in the electromagnetic spectrum, enormous amounts of information could be exchanged without using tons of power.

If they use radio, we could use a big antenna to beam broadcasts at the target star, wasting as little power as possible on surrounding empty space. We could do that today with Arecibo and its one-megawatt transmitter, which is normally used for radar mapping objects in the solar system, although much of the sky is inaccessible to it. If they use optical signals, we could use optical telescopes to send back pulses of laser light, although our technology would probably need to improve considerably. To transmit or receive continuously, we would probably need several sites around the globe, since most stars rise and set as seen by observers on the surface of the Earth.

Any signal from another star would travel through space for many years before we received it, due to the speed of light—possibly hundreds or thousands of years, depending on the distance—and responses would take just as long. Two-way exchanges might not be possible on the time scale of a human life. Time lags are one reason that messages might be rich in information, reducing the need for interaction.

BROADCAST STRATEGIES

What kind of broadcasts might be out there? It's worth speculating about because most search strategies can find only certain kinds of signals. From a searcher's perspective, there are two categories of broadcast—those that are present all of the time, and those that are not.

Continuous transmissions have been assumed by most searches so far, which is an optimistic assumption that allows us to survey the sky fairly quickly (a few years) with a fairly big antenna listening for only a few seconds or minutes in each direction. Such broadcasts could be omnidirectional, radiating in all directions all of the time just like stars and many natural radio sources, or they could be targeted, directed at us with a high-gain antenna system pointing our way all of the time.

Omnidirectional signals require a great deal of power, as we've seen, and assuming so much power means we are presuming a lot about extraterrestrial resources, technology, interests, and so on—so it's open to charges of excessive optimism or wishful thinking.

Targeted transmissions can be much cheaper in terms of power. Someone could illuminate our solar system all the time with a big antenna if they know we are here, perhaps having detected our local radar or television broadcasts or life-gases in our atmosphere—although this also seems optimistic. Broadcasters don't necessarily need to know which stars host potential listeners; they could illuminate many stars simultaneously with many narrow beams, although as the number of beams increases, more of the sky is painted and the power savings decrease.

Intermittent signals might be more attractive to broadcasters because they could be much cheaper in terms of power, making broadcasting possible when it might not be otherwise due to the cost of power. There are several ways that intermittent signals could arise. One is an omnidirectional broadcast that's turned on for only a small part of the time—say for one second out of a hundred, thousand, or million—which would reduce the power bill by the same big factor. Another way is a targeted transmission pointing at a list of stars in succession—say a hundred, thousand, or million stars—illuminating each for only that small fraction of the time. In either case, searchers would need to search in time as well as direction, but they don't know the amount of time between pings. If the signal is too rare, then it's unlikely that a searcher using a single big dish antenna will happen to be looking in the right direction at the right time.

Intermittent broadcasts would be harder to detect using our current search strategies, but might be forced on broadcasters by limited resources. Searchers might cope by using antenna systems capable of monitoring large parts of the sky simultaneously, which is something we can't do very well now with high sensitivity but may be able to do in the future.

Many people think that broadcasters would use a two-part strategy: a strong beacon that's fairly easy to find but sends little information, plus a much weaker data channel that sends information faster but would be harder or impossible to find until the strong beacon attracts the attention of searchers. One reason this scheme seems attractive is that fast-flowing information makes a signal wider in bandwidth, and that makes it harder to detect with narrowband receivers—so if someone wants to be detected, they have a better chance if they supply a narrowband signal although it can't carry much information. If they also want to transfer lots of information, they can do it with a second signal, possibly much weaker and at a much higher frequency that allows more efficient communication. Detecting a beacon presumably would motivate a searcher to look harder at that location, even if it required building larger or specialized receiving gizmos.

Practical considerations might have a big influence on broadcast strategies, such as economics (the scarcity of resources like energy), ecology (avoiding the addition of too much waste heat to the environment), geography (if transmitters are located on rotating planets), and so on.

EXAMPLE BROADCAST STRATEGY

It's unlikely that anyone can correctly guess what interstellar transmitters would look like, but this section presents one scenario that takes into account some of the practical constraints broadcasters might face. It yields a signal somewhat like the Wow detected at Ohio State University in 1977.

Consider the problem of broadcasting from a rotating planet with limited resources—a place rather like the present-day Earth. It's unlikely that we would broadcast in all directions continuously because the power bills would be enormous, and if done from the ground, local folks might beef about being roasted by the heat. Doing it from orbit would add enormous cost and complexity. Using thousands or millions of huge antennas with narrow searchlight beams to direct power toward many stars and constantly track them seems too expensive and complicated.

A much simpler and more economical approach might be to radiate signals in a thin, wedge-shaped beam along a north-south line and let the planet's rotation sweep the beam across the entire sky once each day—call it a rotating fan beam transmission—which would have some advantages of both omnidirectional and targeted broadcasts. One way to do this would be to build roughly one hundred antennas somewhat like Big Ear described earlier, each pointed at a different declination and illuminating an eighth-degree-wide slice of sky along a north-south line, with the Earth's rotation sweeping that cat's eye beam across the sky once each day. Since it would illuminate less than one-thousandth of the sky at any moment, the power bills cost less than one-thousandth of a constant broadcast in all directions, and since the antennas would not need to move, they could operate for a long time without mechanical problems. The scheme would offer the full-sky coverage of an omnidirectional broadcast, although not full-time, and offer some of the power economy of directional antennas.

A distant observer would see this rotating fan-beam as a flash lasting about half a minute and repeating at 24-hour intervals, like a lighthouse. Finding the periodic flash for the first time would be a challenge, but searchers have a hint as to how long to look because they probably know something about the length of planetary days. Planetary days can't be much shorter than six hours because rocky planets can't spin much faster without breaking up, and planets with days many times longer than 24 hours might have climates inhospitable to life due to extreme temperatures. In our solar system, four planets have days of between 10 and 17 hours, and two others are between 24 and 25 hours—six out of our eight or nine planets have days of 25 hours or less (the XXVI General Assembly of the International Astronomical Union meeting in Prague voted in 2006 to adopt a definition of "planet" that excluded Pluto, leaving eight). If planets in other systems have a similar range of periods, searchers could have a rough idea of how long to look in any direction for such beacons.

This hypothetical broadcast strategy is intended to show one way periodic or intermittent signal broadcasts might arise, but it's only one of many broadcast strategies that might be out there. We don't know what to expect, but spinning planets may play a role.

SUMMARY

Interstellar communication is definitely doable and might be fairly easy once both parties know each other's location and wavelength. The big challenge is first contact, either by guessing unknowns like frequency and perhaps location, or if that fails, by a systematic and maybe lengthy search.

Searchers have so far sought easy pickings: an advanced civilization sending a strong signal that's present all the time, so it's detectable with our relatively low level of effort—typically a professor working part time with a modest budget. That seems like a sensible starting strategy, but the disparity of effort is striking. It's possible that no broadcasters dedicate vast resources to getting to know their neighbors, and that we need to look for broadcasts being made on the economy plan.

Search strategies are at the core of SETI, since it's simply not possible for a search to cover all directions, wavelengths, signal strengths, time periods, and other possibilities.

CHAPTER 11

SEARCH STRATEGIES

COSMIC HAYSTACK

Searching for signals from the stars has been compared to looking for a needle in a haystack, but it's much harder. We don't know what direction signals might come from, their frequency or strength, if they are "on" all the time, or anything else about them. We don't know what the needle looks like or if it exists at all! Covering all of the possibilities is impossible; there could always be a weaker signal or higher frequency than a practical search can detect, as illustrated in Figure 11.1.

Search strategies select some part of the cosmic haystack that can be searched in a reasonable amount of time, hoping that a clever enough strategy will find a signal sooner rather than later. Examples include targeting older stars where life and intelligence have had more time to evolve, or selecting bands of frequency that seem likely to be used for broadcasts, or listening for extended periods of time for intermittent signals.

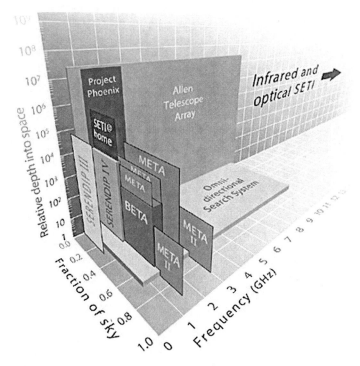

Figure 11.1. Three dimensions of the cosmic haystack—frequency, direction, and sensitivity—and the modest portions of that search space covered by major radio searches to date. Illustration courtesy of *Sky & Telescope*.

THE PROBLEM OF FREQUENCY

Sifting through even small parts of the vast electromagnetic spectrum is one of the biggest challenges in SETI. Covering the terrestrial microwave window from 1-10 GHz with 1-Hz resolution would take nearly ten billion channels, exceeding the biggest spectrum analyzer built so far. Searching the free-space window up to roughly 60 GHz would require even more channels and observing from space as well because some parts of the spectrum are absorbed by the atmosphere. There is much more spectrum extending up to optical wavelengths, and much more beyond, such as x-rays, as shown in Figure 11.2.

"Magic" frequencies offer a possible solution to the problem of frequency, hoping that broadcasters make signals easier to find by choosing a frequency or band that searchers can guess, or that astronomers might stumble across in the course of their work. Some have suggested the 21-centimeter hydrogen line, and many searches have been done near that 1.42 GHz frequency. But not everyone agrees; other magic frequencies have been suggested, and we don't know what might seem sensible to alien minds. Calling them magic is a bow to the fact that they might not be.

Multi-channel spectrum analyzers (MCSA in acronym-ese) are currently the best tool for radio searches, making it possible to monitor many frequencies at once. The biggest analyzers to date were Harvard's BETA, with 240 million channels covering the band of frequency from 1.4 to 1.7 GHz in eight steps, and SERENDIP systems with up to 168 million channels.

Digital processing power has doubled roughly every two years for decades, so the capacity of digital spectrum analyzers has increased with incredible speed. Harvard's 240 million channel BETA analyzer of 1998, for example, had thousands of times the 65,536 channels of the 1978 "Suitcase SETI" analyzer (named for its size). A ten-billion channel analyzer able to cover the whole terrestrial microwave window is easily conceivable today although not yet easily affordable.

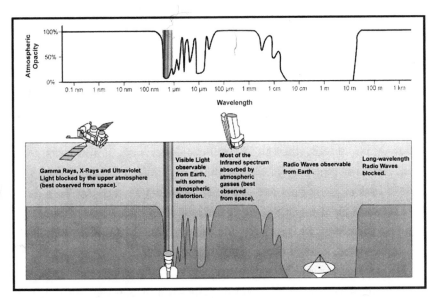

Figure 11.2. The electromagnetic spectrum and opacity of our atmosphere at various wavelengths, with short wavelengths like gamma rays and light toward the left side, and longer wavelengths like radio toward the right. Source: NASA.

A similar problem of wavelength selection exists with searches in the optical realm; it would take a long time to comb through many different wavelengths looking for a slight excess of light of one color that might be due to a laser aimed at us from near a star. Some optical searches assume a laser that's strong enough to outshine its home starlight at all wavelengths, and that seems possible if the flashes are very brief, although we need to make billions of measurements per second to catch such quick flickers.

THE PROBLEM OF DIRECTION

Big radio and optical telescopes "see" only tiny spots on the sky. Arecibo, as noted earlier, sees only about one ten-millionth of the sky at any one time—a spot about one percent of the area covered by the Moon. With such small fields of view and limited time, we have to decide which directions to observe. That means we might be looking in the wrong direction and miss signals arriving from elsewhere, or we might be looking in the right direction but not be looking at the right time when an intermittent signal appears.

Targeted searches exploit the directionality of large antennas by using their high gain to get high sensitivity on selected targets. These searches often select stars that are Sun-like—old enough to have evolved intelligent life based on what we know of stellar, planetary, and biological evolution—a bet that could pay off if a signal is present when we happen to look at a target star. But there are lots of possible problems. One is that the selection rules might be wrong, leading us to look at the wrong kind of stars. Another problem is that we might look when an intermittent signal is off, or a when targeted broadcast is not targeting us. Perhaps the worst problem is that it takes a long time to look at many stars one-at-a-time. Dwelling on each star for a thousand seconds, for example, means we can look at only a few tens of thousands in a year (assuming we have only one beam); if one in a million stars has a detectable signal, it could take a generation to find it.

Sky surveys are an alternative strategy, typically using wider beams to cover many stars at once. If even one sufficiently strong signal is out there and present all the time, then an all-sky survey should find it, and antennas in the 30-meter range can survey the sky in a few years. One drawback is lower sensitivity due to smaller antennas and shorter dwell times—typically one-hundredth or one-thousandth of the sensitivity of targeted searches, which often makes surveys shallow. Another drawback is less chance of catching intermittent signals, because the beam typically sweeps over each area for only seconds or minutes rather than dwelling for a while.

Isotropic searches monitoring the full sky full time might seem like a logical solution to the problem of direction, but with current technology and relatively modest funding, such searches are not very sensitive. Watching the forest rather than looking at one tree at a time would be appealing for both SETI and general radio astronomy and may become feasible in the future.

All ground-based searches have the problem that no single site can see both the entire northern and southern skies. Only one radio search has surveyed the entire sky—Harvard's META in the Northern Hemisphere and the Argentine META II in the Southern.

SEARCHING GALAXIES

One strategy for dealing with the problem of direction is to search parts of the sky with the most stars, so we can search many simultaneously. Stars are concentrated in the disk of our galaxy, so looking along that glowing Milky Way is one way to see more stars in each glance; looking toward the center sees even more stars at once.

Other galaxies offer even larger concentrations of stars, letting us look at billions at the same time. That eliminates some unknowns, since both sides know where to point their antennas and know the range. Another attraction is that we need not assume our Sun is a target for broadcasts, merely that our galaxy is—and that's no great conceit because the Milky Way is one of the two largest galaxies in the local group of a few dozen.

Carl Sagan and Frank Drake did a quick search of the Triangulum galaxy M33 in 1974, a spiral like our own some three million light years away, using the Arecibo antenna. Since its beam is much smaller than that several-degree galaxy, they listened only briefly at each of many different spots. No luck.

I did much longer galaxy searches in 1989 and 1991, using the META system to search both M33 and the Andromeda galaxy M31, which have tens of billions and one thousand billion stars, respectively. The antenna's beamwidth was one half-degree, smaller than the several-degree angular size of the galaxies, so a half-dozen areas were tracked for five hours each. No big peaks were seen at M31, but several were seen while observing the center of M33— more or less coinciding with the arrival of an automobile at the observatory, and not seen again.

Spanning the vast distance between galaxies requires lots of power. With Cyclops-scale 3,000-meter antennas on each end, a single 1,000 MW power plant could provide enough juice to signal over the roughly million light years to nearby galaxies—but searching with such narrow beams resurrects the problem of where to look. If we search with a system like Harvard's META—able to view much of a galaxy at one time—then we could detect a similar-scale antenna illuminating the Milky Way if it broadcast 10^{17} watts. That's a thousand times terrestrial production, but requires only one power-rich civilization in a galaxy. Broadcasters could reduce the power bill by a factor of 10,000 by using a Cyclops-scale antenna, although it would illuminate only part of our galaxy at any one time so we would need to listen longer.

One attraction of searching galaxies is that we can listen for days, weeks, or longer to catch intermittent signals from billions of stars.

ISOTROPICALLY DETECTABLE BEACONS

In everyday life, we rarely care where transmitters are located and rarely use huge antennas to receive them; broadcasters supply enough power to be detected by small antennas in our radios and other gizmos that pick up broadcasts from all directions (satellite TV is one exception, but those antennas are shrinking). If interstellar broadcasters want to be detected, they might provide signals strong enough to receive with omnidirectional antennas, so that searchers need not worry about where to look.

A signal that's "strong enough" could be called an isotropically detectable flux. We could calculate how strong that would need to be for a truly omni-directional receiving antenna, but such antennas would not be very practical because they pick up noise from the ground and local interference. Assuming instead small one-meter antennas, roughly a hundred could cover the entire sky, and a detectable signal would be roughly 10^{-20} W/m^2 at 21 cm. To deliver such a strong signal, broadcasters would need to use huge amounts of power or huge antenna systems or some compromise between the two.

Radiating in all directions constantly would require 10^{19} watts to be detected at 1,000 light years by a one-meter antenna—a thousand times the solar energy falling on the Earth, which seems extravagant. Broadcasters could use really big antennas to save power, in an extreme case, by illuminating only the area with planets near a target star. That would reduce the power to as little as 10^6 watts per target or one-thousandth of a big power plant, regardless of distance. Forming such needle beams at the hydrogen frequency would take an antenna system 15 kilometers across for a range of 100 light years and ten times larger for 1,000 light years, although using higher frequencies would allow smaller antennas. Such big antenna systems seem extreme, but we don't need to build them; the burden falls on broadcasters if they wish to maximize their chances of being detected.

One interesting aspect of huge antenna systems is the possibility of using the large area to generate power with solar cells, possibly making a broadcast energy self-sufficient. Or it might work the other way around—sprawling solar power collection systems might have tiny transmitting circuits built in taping a bit of power. Or transmitter elements might be integrated into zillions of scattered devices—cell phones, for example, coordinated by "apps" and broadcasting while the devices are charging.

Signals strong enough to be detected with very small receiver antennas might be intermittent, since the very big transmitting antenna systems needed to form needle beams suggest a targeted broadcast strategy with one or perhaps many beams moving from star to star. Searching for intermittent signals would require monitoring large areas of sky for extended periods of time, but that's fairly easy because the assumption of strong signals allows a modest number of small antennas to monitor the full sky full time.

THE PROBLEM OF TIME

Most searches assume a signal will be present whenever we happen to look—just like stars shine all of the time—but signals might be intermittent. If so, most of our surveys would fail, which could be one reason that we have come up empty-handed so far.

This problem can arise from an omnidirectional broadcast being "on" only part of the time, or a targeted broadcast not targeting us all of the time. The solution would be to monitor much or all of the sky continuously, but that's hard because detecting weak signals usually depends on a telescope gathering radiation from one direction to the exclusion of all others. It's possible to imagine having many big telescopes monitoring many different targets simultaneously, but that scheme gets expensive.

In the radio realm there are several techniques that promise the ability to monitor large areas of sky, with the Allen Telescope Array making a start at monitoring five square degrees simultaneously, and the Argus array under development at Ohio State monitoring much of the sky continuously although at low sensitivity, both discussed later.

The problem of time is especially tough because there is no obvious magic interval that might be used to determine the time between signals. Signals at intervals of seconds or minutes might be detected by many of our searches, but signals at intervals of hours or longer probably would not be. Much longer intervals like months or years would almost certainly be missed. One rough natural period might be the scale of a typical planetary day, which is probably in the range of tens of hours if our solar system is a reasonable sample.

OPTICAL SEARCHES

Broadcasting with light has the glaring problem that starlight from the broadcaster's nearby star would tend to obscure the signal because it takes a mighty bright light to outshine a star. But the idea has attractions too; the optical telescopes needed for searching can be much smaller than the antennas needed for radio, and our most powerful lasers today can briefly outshine stars.

The first working laser was built in 1960 by Theodore Maiman. Arthur Schawlow and Charles Townes had developed the idea, filing a patent in 1958 and eventually receiving Nobel prizes (Townes in 1964, shared with A. Prokhorov and N. Basov; Schawlow in 1981). Townes wrote a paper in 1961 proposing optical signals for interstellar communication, but it took decades to develop lasers strong enough to make optical broadcasts seem plausible. Our most powerful lasers currently produce one megawatt continuously, and a petawatt (10^{15} watts) briefly in picosecond pulses (10^{-12} second, or 1 trillionth), which might do the job if directed by a big telescope.

Optical searches typically look for pulses of light outshining a star for a tiny fraction of a second. In an example crafted by Paul Horowitz, about one million (10^6) photons of light rain down on each square meter of the Earth every second from a star 1,000 light years away. If we intercept those photons with a telescope having a square meter of collecting area and measure how many arrive every microsecond (10^{-6}), we would see an average of one count in each measurement. If we measure much faster, say every nanosecond (10^{-9}), then most of the billion samples during one second would be completely free of photons from the target star—the starlight essentially goes away except for occasional single counts. A bunch of photons arriving during a single nanosecond would be surprising.

Such powerful pulses of light could be produced by lasers comparable to the Helios system being built at Lawrence Livermore National Laboratory's National Ignition Facility, which is capable of producing several megajoules of energy for a few billionths of a second (one joule is the energy required to exert a force of one newton for a distance of one meter, or more tangibly and amusingly, to lift a smallish apple over your head). Beamed through a 10-meter telescope to illuminate just the area around a star 1,000 light years away where habitable planets might orbit, such a laser could deliver a flash of roughly 500 photons per square meter for one nanosecond—far more photons than expected from starlight alone. Optical SETI looks for such pulses, which would be a strong sign of intelligence and could carry information by their timing.

The cost of this sort of signaling depends considerably on how many pulses are sent. Using rough numbers, sending one pulse of one megajoule (10^6 joules) toward a single target would cost a few pennies, which seems incredibly cheap (3.6×10^6 joules makes a kilowatt-hour, and power companies sell kilowatt-hours for roughly ten cents). But if an optical beacon was used to ping a million target starts once each second, to take an example roughly comparable to an isotropic radio broadcast, it might cost $100 million per hour, which seems steep.

We could detect such optical pulses today if they repeated often, although a practical optical broadcasting system is beyond our current technology. The Helios system, for example, was designed to produce only a few powerful pulses each day. Also, illuminating just the planetary zone around stars—the main advantage of optical transmissions—requires knowing the star's motion with extreme accuracy in order to point the beam where the star *will* be when the signal arrives many years in the future. Practical broadcasts would probably need to illuminate many targets, and pointing a laser beam focused by a big telescope at many stars quickly seems challenging.

The bad news about optical communication is... weather. Imagine a brave band of photons marching across the lonely void between stars for centuries or millennia, stoically bearing their message. If they're of the optical persuasion and happen to arrive on a cloudy night, they tragically die just before reaching ground-based telescopes awaiting their message. If they're microwave photons, though, they break through the clouds unscathed and deliver their message to a (hopefully) waiting radio telescope.

Clouds are probably not just a terrestrial problem; other warm, wet worlds would likely have them, too. We put telescopes on high places like mountains and volcanic peaks in part to get them above the clouds, and some little green astronomers elsewhere might do the same if they have the right geography. Satellites, of course, could carry telescopes above the weather, but orbit is a very expensive neighborhood. Putting things at the mercy of weather seems like a handicap, at least for initial contact.

Another optical problem is extinction—dimming caused by the dust and gas in interstellar space—which makes much of the galactic plane like a brick wall to light over long distances. Many radio frequencies, however, pass through the interstellar medium like light through clear glass.

Ambitious optical searches are in progress or planned at Harvard, Princeton, Berkeley, and University of Western Sydney, Australia.

EAVESDROPPING

What if nobody's broadcasting? There could be gizmo-builders galore out there, but if none send signals intended to span the stars, finding them would be hard. The nobody-broadcasting scenario might not be fatal, though, because we could try eavesdropping on transmissions used for local purposes—detecting leakage of local signals into space.

It's hard to guess what sort of signals might be used for local purposes elsewhere, and many would probably not be strong enough to detect with our current technology. On Earth, the trend is away from broadcasting lots of power "on the air", and instead sending signals through wires or fibers or beaming them down from low-power satellites. Cellular phones, for example, broadcast very little power (around one watt) which allows many people to use the same frequency in neighboring areas without interference. Calls typically travel over the plain old telephone service for most of their trip (POTS is the inelegant acronym, originally from post office telephone service). More advanced civilizations might radiate less power for local communication.

Radar is one type of transmission that must go out over the air and must be fairly powerful, and it seems likely to be used wherever aircraft fly or folks want warning of approaching comets, missiles, or other worries overhead. Radars radiate a lot of power because only a little bit hits a target, and what bounces back to the receiver is weaker yet (falling with the inverse fourth power of range). We can try to guess something about the kind of radar signals ET might use. First, radio is probably used rather than optical because of the cloud problem. Second, very high frequencies probably aren't used, at least on wet planets, because they are absorbed or scattered by water vapor and rain. Third, frequencies below microwave are probably not used because the antennas needed are too big. The band of frequencies remaining is roughly the same terrestrial microwave window that already looks promising for SETI. One problem with looking for distant radars is that radar uses very brief pulses of power, which appear as wideband rather than narrow signals, and would be hard to detect with the receivers using narrowband channels that our present-day searches typically employ.

Radar is the most easily detectable sign of life on Earth; Woodruff Sullivan and others have shown that our ballistic missile warning systems are potentially detectable from nearby stars. Very powerful radar signals from other stars might be detected with our current large antennas, but serious eavesdropping might require much bigger antennas.

ARCHIVAL SEARCHES

Radio and optical astronomers have generated hundreds of catalogs and millions of images of the sky, and it's conceivable that some sign of interstellar signals lurk in those archives. Most of this stuff has already been analyzed, but astronomers are usually not looking for intelligent signals. In fact, the first thing most radio astronomers do with raw data is to *throw out any radio signals*, assuming that they are man-made interference. The original data is usually retained, with the apparent interference simply flagged and ignored, so archival searches could review such events.

We can also cross-reference existing catalogs and images looking for potentially interesting features. One example is looking for stars that appear in both radio and optical catalogs. Few stars have detectable radio emission, so finding a radio source at the position of a star is potentially interesting. Figure 11.3 shows a bright star found by matching the positions of radio sources in the FIRST survey (Faint Images of the Radio Sky at Twenty centimeters) with the Guide Star Catalog, which lists positions of over ten million stars and galaxies. Cross-referencing the position to the SAO (Smithsonian Astronomical Observatory) catalog, the star turns out to be of spectral type G5, which is close to the Sun's G2 type and in the range considered promising for the evolution of life and intelligence.

The radio source does not appear in the NVSS (NRAO VLA Sky Survey) radio survey covering the same area and frequency, although it should have

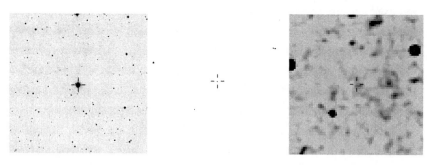

Figure 11.3. The bright star (SAO80232) in the image on the left has the same position as a weak (5 mJy) radio source shown in the crosshair of the center panel found by the FIRST survey, yet does not appear at the same position in the crosshair in the NVSS radio survey shown in the right panel. Sources: Digitized Sky Survey, FIRST, and NVSS.

been detected by both surveys. Its absence in one catalog suggested a variable flux, which could be due to natural causes but conceivably could be due to a signal present during one observation and not during the other. Checks of yet other catalogs found the star listed as a strong source of x-rays, which would be expected of a radio star, although they are rare, and not of the home star for an interstellar broadcast.

This example of an archival search was a bust, but it illustrates how the growing body of astronomical information might be used to find interesting things.

GRAVITATIONAL LENSING

Distance is the bane of astronomy and SETI because electromagnetic radiation spreads out as it propagates and is very weak by the time it reaches us. Ever-larger telescopes are built to see further and fainter objects, but there are limits to the size of things we can build. One exotic way to simulate a titanic telescope would be to use the Sun like a huge lens.

A star's gravitational field bends light from distant sources as the light passes by, concentrating it along a line radiating away from the star in the opposite direction. Looking back toward the Sun from the right distance, it's thought that objects located on the far side of the Sun would appear very much intensified. The effect has been observed by astronomers, who sometimes see distant objects like quasars appear as multiple images next to intervening galaxies. It's a striking notion that every star is surrounded by the amplified images of everything in the distance.

Since the effect also works for radio waves, some have suggested sending radio telescopes out to the magic distance to listen. Unfortunately, the rays do not begin to converge until roughly 550 astronomical units past the Sun, which is more than ten times further out than Pluto (an astronomical unit is the average Earth-Sun distance). One problem is traveling out far enough to observe the effect, and another problem is moving from one point of interest to another in the focal region to look in various directions. We may one day dispatch probes to exploit gravitational lensing but probably not very soon.

Astroengineering

Physicist Freeman Dyson suggested that some large-scale engineering projects of advanced civilizations might be detectable as warm spots on the sky. A civilization could, he argued, harness a large fraction of its star's energy by enclosing it inside a shell, and such a Dyson sphere might be detectable by infrared telescopes because it would re-radiate the energy of the star at a temperature of only a few hundred degrees rather than the much higher temperature of the star inside.

Dyson was thinking about the limits to growth, because many years of just a few percent annual growth could increase population and presumably energy needs beyond the capacity of planets. Only a tiny fraction of a star's energy falls on planets (roughly one-billionth hits the Earth), but ambitious folks might intercept more. The upper limit to the amount of material available for collecting solar energy and for suburban sprawl would be the total mass of a planetary system—which Dyson suggested could be used to construct a shell surrounding the star. An actual shell would be impossibly flimsy and unstable, but other approaches, such as a zillion orbiting habitats, might work; physicists like to leave the details to engineers, who will hopefully remember to deal with comets and other inconveniences.

The idea that waste heat from large-scale space engineering projects might be detectable has led to several searches that have not yielded any exciting results so far. Exponential population growth that might motivate such projects is often countered by factors like lower fertility or higher mortality, and stable or declining population is now seen in some developed nations. But it seems a smart bet to consider the possibility that intelligent life might extend into space and that its activities might be detectable.

Other Signs of Life

It is becoming possible to identify some gases in the atmospheres of some planets orbiting other stars, and gases such as ozone (O_3 in chemistry shorthand, a short-lived variant of the O_2 oxygen molecule) could suggest the presence of life. The oxygen in our atmosphere is constantly replenished by photosynthesis of plants and other organisms; without them, it's thought that most free oxygen would react with other stuff—rusting iron, for example— and largely disappear. Planets with life but little or no oxygen, like the Earth for several billion years, might be marked by other gases such methane. Spectra for several of our own planets are shown in Figure 11.4.

One method for finding planets and "seeing" them is to watch for brief dips in the brightness of stars due to the transit of a planet across it. This works only if the planet's orbit happens to be in the right plane as seen from Earth—odds are less than 1% for planets in Earth-like orbits—and it happens only one time each planetary year. During a planet's transit, we can study the spectrum of the starlight passing through its atmosphere to learn something about its composition which might reveal signs of life.

Another way to see a dim planet near the glare of a nearby star is nulling interferometry, where the light from two or more telescopes is combined so that the crests and troughs of the star's light cancel, while the light from a planet slightly separated from the nulled point does not. Yet another method uses a coronagraph, which blocks the star's light so that faint nearby planets can be seen. Either approach would require large space telescopes, possibly orbiting far out in the solar system.

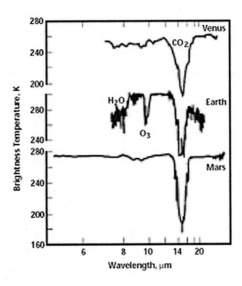

Figure 11.4. Spectra of Venus, Earth, and Mars showing absorption by gases. Carbon dioxide is present in all three; ozone and water vapor appear only on Earth. Source: NASA.

Finding signs of life in the atmospheres of distant planets would be an intriguing discovery, although it would not quite prove life was there because we might not be able to rule out all other possible explanations. Targeted searches might scrutinize such stars for signs of technology, and over time we might come to monitor many promising planets.

Signs of life might also be found in the spectra of stars if they were altered by local residents. Folks might dump artificial elements with short half-lives into their stars to make their spectra show signs of intelligent activity, or might dump radioactive waste into the star to get rid of it. The presence or over-abundance of elements like technetium, plutonium, or others could mark a star as having something fishy going on.

Viewing planetary surfaces from orbit or spacecraft on the ground is another way we might find life. We can already look down on the planets and moons of our own solar system from orbiting robotic spacecraft, and on a hundred- or thousand-year time scale we might do the same for nearby stars.

SUMMARY

Search strategies aim to find signals by using what we hope are smart schemes to reduce the huge number of possible frequencies and places and times that we might otherwise need to examine.

So far, not many different search strategies have been tried—there are more ideas and possibilities than there are people and funds to put into action. Radio searches have covered only small parts of the spectrum and have not listened at many stars for very long; optical searches have only recently begun in earnest.

As our instruments get better, we will also be able to search for unintentional signs of life, like atmospheric gases caused by biology, and to spot planets orbiting their stars in habitable zones.

With smart strategies and a little luck, we might find ET before too much more time passes.

CHAPTER 12

CONTACT

Detecting just one interstellar signal would prove the existence of extraterrestrial life, intelligence, and technology—the whole enchilada—even if we couldn't understand it. Discovering Others would be historic and profound, although after lots of initial media razzle-dazzle it might not make much difference in everyday life. This chapter speculates on how a detection might occur, possible contents of messages, possible consequences, and thoughts on responding—but with no money-back guarantee because as Niels Bohr reportedly said quoting an old Danish proverb, "It is difficult to predict, especially the future."

EUREKA!

There's a good chance that nobody would hear or see a signal when it was first detected because searches are usually automated. The popular picture of stunned scientists staring at a signal surging up on a computer screen is one possibility, but events might unfold much less dramatically—with a signal being found only after lengthy analysis and additional observation because it might be very weak (having traveled far), ambiguous (not well-matched to the receiver), and hard to reproduce (maybe not present all of the time). Days, weeks, or much more time might pass before researchers were sure it was the real thing, and most scientists would want to know a lot about a strange signal before claiming to have discovered aliens.

If such an announcement ever occurs, it will probably be made by a group representing several observatories who have studied a signal enough to be sure it's the real deal, and who have prepared themselves to answer a barrage of questions from a mob of reporters and skeptical colleagues.

Guidelines for announcing a discovery have been circulated by the International Academy of Astronautics under the lengthy title *Declaration of Principles Concerning Activities Following the Detection of Extraterrestrial Intelligence*. Those pages recommend that one be sure a signal is extraterrestrial and artificial, get independent confirmation from another observatory, then announce to astronomers, the public, and a long list of organizations (which presumably might otherwise overlook the news).

Some folks worry that the news would be suppressed by the Authorities— or that it already has been—but that seems unlikely because news travels fast in the scientific community, and bragging rights for discoveries are very highly valued.

GETTING THE MESSAGE

We would want to know what message a signal carried (assuming there was any), but most instruments used for searching are designed only to detect the presence of signals rather than extract information. Once a signal was found, we would probably need to customize receivers to get a better record for study, and actually "getting the message" would have several additional steps, including demodulation, deciphering, and comprehension.

Modulation is the process of encoding information onto a carrier signal, which the sender does and the recipient needs to figure out—it's quite like what we do with our voices when we talk or sing. In radio, the carrier is usually an electromagnetic wave with a fixed frequency like a pure tone, and in the simplest kind of modulation known as continuous wave, the carrier signal is simply switched on and off as in Morse code. A carrier can also be modulated to carry information like voice, music, video, and so on by varying its amplitude (AM), its frequency (FM), or some other property such as polarization. Many searches anticipate signals switching polarization between left and right circular, which seems nicely matched to the binary coding we use in the present digital era.

Demodulation is the process of extracting information from a signal, which the recipient needs to do. It's fairly simple once you recognize the type of modulation used—a diode and a few other electronic parts do the trick for AM radio. Very weak or fast-changing signals might require bigger antennas or specialized receivers to make out the details, but we could probably manage it without too much trouble.

Once we had the raw information—say, a sequence of beeps—we would need to decipher it into a form we could recognize. That might be tough because given a bunch of beeps, how do you decide what they represent? If they came from your shortwave receiver, you might recognize certain sequences as letters of an alphabet in Morse code, and with luck, the letters might form words in a language you understand. But what if the beeps represent music, images, computer programs, or something else? It might be much tougher for extraterrestrial messages, because broadcasters are unlikely to use any code or language we know and might be sending something we would not recognize like (heaven help us) coded odors. Presumably such broadcasts would be designed to make deciphering as easy as possible.

Comprehending a message is likely to be the big challenge. If a message consisted of word-like symbols, matching them to some words in our languages might be fairly easy for shared concepts once we spotted them, such as "the number one," "star," and so on. The fact that technology was used to send the message would guarantee some common ground, but there would surely be many concepts we don't share.

Comprehension might be easiest in mathematics and science because the properties of numbers, matter, and energy are universal, and the symbols used to talk about those topics can be simple and logical (although not all terrestrial students would agree). For example, symbols like .~. ..~.. ...~... might suggest that "~" stands for "equals," that groups of dots stand for numbers, and that groups of symbols are statements or equations. Much can be communicated this way; continuing the example, a symbol for addition could be defined by a sequence of symbols like .|.~.. .|..~... ..|.~... and so on. The mathematician Hans Freudenthal developed the language Lincos (for *lingua cosmica)* along these lines in 1960.

Differences between broadcasters and searchers would complicate understanding. Modern humans can translate between languages because we are talking about the same things regardless of the words used—not only with external objects, but with internal affairs like feelings—"love" and "l'amoure," for example, refer to the same emotion in both English and French. But translation fails in the absence of common ground, even for humans. How could "digital computer" be translated for a cave man lacking knowledge of either machines or calculations? In the case of interstellar contact, we are the guys wearing animal skins (as belts, shoes, or jackets; occasionally fur hats or coats). The situation is likely to be worse between species and worse yet if the message makers have major senses differing from ours, based on bioluminescence, echolocation, electric fields, odor, or many other possibilities.

Many problems might be avoided if messages included images; given pictures, we could figure out a lot for ourselves. Images would let recipients comprehend something of other worlds in their own terms, letting them see exotic environments, strange creatures, and indescribable things. That would require vision somewhat like ours, but it need not be identical; we use our eyes to see many things we can't see unaided, assisted by tools such as radio telescopes, infrared cameras, and X-ray machines. Images would be more expensive to send than words because they contain more information, but potentially deliver much more bang for the buck.

An interstellar message might be tough to comprehend, but it's hard to imagine that we would not be able to puzzle out enough to find it interesting.

WHAT MIGHT WE FIND?

Some scientists have suggested that an initial signal might prove its intelligence by way of mathematics—perhaps consisting of a string of prime numbers (beep-beep, beep-beep-beep, beep-beep-beep-beep-beep, and so on), which nothing other than intelligence is expected to produce. Others have toyed with crude pictograms showing the shape of the sender, number of planets in our solar system, and so on. Beacons intended only to draw attention might carry such simple information—or even none at all, other than announcing its existence—but hopefully would be drawing attention to a signal with much richer content.

An enormous variety of stuff can be transmitted, including simple messages like "Are you blokes Episcopalians?", books and entire libraries, sounds, music, pictures, movies, computer programs, and more. Pleasantly, transmitting such things does not reduce the sender's supply of them.

What might motivate ET to broadcast, anyway?

Trade in knowledge might motivate some broadcasting between stars, just as trade in goods motivates much terrestrial broadcasting. Many kinds of knowledge can be stated in simple equations that have enormous value when understood and put to use—Einstein's famous $E=mc^2$, for example, is worth a fortune if you don't already know it, since it reveals a way to generate energy from rocks (after extracting uranium) and it explains why stars shine.

Trade in descriptions of other worlds might also motivate broadcasts, since it would be much more convenient to have local folks show and tell about their surroundings than it would be to visit. Alien *Wild Kingdom* programs might be fascinating, and we might send ours in return. Trade in culture might also motivate broadcasts. Human culture is a vast collection of stuff unavailable anywhere else, and some of it is highly valued in spite of differences like language and culture. The Mona Lisa's smile and Playboy Playmates' charms are probably too subtle to intrigue other species, but Picasso's cubist paintings might be an alien sensation. Our history might be interesting to folks elsewhere who want to know how others arose, and the histories of other worlds might be valuable to us; it would be worth knowing if there are paths that always lead those venturing down them to gloomy woods and ruin, or royal roads that really do lead to Utopia.

Exchanges would be slow due to the speed of light, and since responses might take a very long time, initial messages might contain a lot of information to serve as a menu and teaser, or perhaps the entire story.

Some broadcasting doesn't depend on trade, such as stations operated by governments and some educational and religious programming. The same things that motivate humans to speak their minds to the world might motivate others to spread their thoughts across the stars. We might find something like an encyclopedia of another world, and it's possible that we might find an *Encyclopedia Galactica* containing the collected cultures of many worlds. And it could be even better: something like an Internet/TV/Library of Congress fire hose of information and imagery.

The cost of broadcasting would probably constrain the amount and type of material sent, but it could be very inexpensive for nearby stars if two big antennas or telescopes point toward each other. For stars in our 100-light-year neighborhood and Arecibo-scale antennas on both ends, the power required is comparable to a television station: 100 kilowatts, which costs roughly $10 per hour.

IMPACT OF CONTACT

What impact would interstellar contact have?

Science fiction often shows conflict resulting from contact, but strife is a writer's tool and fiction's goal is to entertain, so sci-fi seems a poor guide. People have predicted everything from Disaster to Utopia. Pessimistic views include invasion, culture shock, depression if our smarts turn out to be second-rate, loss of initiative from getting easy information, Trojan horse software or machines, being casually exterminated, and worse. Optimistic views include receiving vastly advanced science and technology, learning our true place in the scheme of things, taking a giant step toward Utopia, joining our fellow beings in a Galactic Club, and so on.

Human history provides only rough guides. When different cultures have come into contact, the consequences have ranged from bad (warfare, subjugation, and disease) to good (trade, exchange of knowledge and culture), and often some of each. The bad usually involved conflict over tangible things like land, treasure, or other resources, but no physical contact is expected from interstellar contact. Merely *knowing* that other groups of humans exist or *receiving information* from them does not seem to have been a major problem in our history.

One often-cited example of information contact is the impact of classical Greek knowledge and ideas. They spread throughout the Mediterranean world thousands of years ago, sparking advances in many places. Reaching the Arabic world, they stimulated advances in mathematics and astronomy, causing no obvious harm. Lost to Europe during the Dark Ages, the re-intro- duction of Greek books and ideas from the Arabic world sparked a renais- sance in the twelfth century that laid the foundation for today's civilization. Armed conflict did occur between European and Arabic empires, but was not caused by the exchange of Greek information in any obvious way.

An example of how contact might go wrong is the cargo cults practiced for a time by small groups of people in parts of New Guinea after World War II. Seeing manufactured goods delivered by ship to harbors and by aircraft to runways, some Melanesians built their own mock docks and ersatz airstrips and performed ceremonies to attract the delivery of their own goods to begin a Golden Age. Unsuccessfully. It seems unlikely that many people would react to the discovery of extraterrestrials in that way, but a few have prepared landing sites to attract flying saucers. Unsuccessfully.

Many people—a majority of Americans in some surveys—believe that life is likely on other worlds and would probably not be overwhelmed by news of its discovery. In a wildly optimistic scenario, we might learn how to get easy energy, antigravity, immortality, and even better television programming, but the goodies would be unlikely to appear overnight and might not come cheap. In pessimistic scenarios (no goodies; maybe bad stuff), it would be worth knowing that there are other thinking creatures out there.

WHAT WOULD THEY LOOK LIKE?

It's a pretty safe bet that intelligent extraterrestrials would not look like humans with pointy ears or Amazon warrior ladies, as sometimes depicted in movies and television.

Life comes in an enormous variety of shapes and sizes on Earth, and dif- ferences on other planets such as gravity, temperature, and more would prob- ably brew up life in forms we can't imagine. But it's fun to try! Carl Sagan, for example, imagined living gasbag "floaters" and other creatures in the dense atmosphere of Jupiter-like planets.

People have speculated that smart creatures elsewhere would probably have organs like eyes to perceive their surroundings, preferably two or more to gauge distance, likely located on top to see further, plus fingers or ten-tacle thingies to manipulate stuff. This line of thinking leads to critters a lot like terrestrial ones—rather like us—and might prove to be correct in some cases. But it's easy to imagine no eyes (deep underwater or underground), no fancy manipulator thingies (like fish), and many other differences that result in creatures very unlike us, and perhaps unlike anything on Earth.

A few broad constraints might limit the size and shape of intelligent land-dwellers on Earth-like planets. Intelligent critters probably are not very tiny, if trillions of interconnected nerve-like cells are needed to be smart (although social insects might be another path to something like intelligence; termites build skyscrapers, relative to their size). Neither are they likely to be immense because of problems like structural strength and regulation of body tempera-ture (our largest creatures are aquatic). Bipeds with heads on top might not be extremely tall because if they tip over braincases might break (our tallest animals are four-legged).

Some aliens might not look like living creatures at all—they might be sen-tient software. This possibility is worth considering because smart critters might create artificial intelligence in computer-like gizmos or copy their own minds into machines. Such creations might outlast the creators, so machine or otherwise inanimate "civilizations" might over time come to outnumber biological ones.

Any interstellar signal we detect is likely to be from a party more advanced than we are and possibly smarter. The reason to think so is that broadcast-ers could be millions or even billions of years older than we are, and if more evolution can yield more intelligence, they could be further up on the scale of smarts. There could be lots of folks out there dumber than us, but they aren't likely to be broadcasting and our searches won't find them. Since we are new to the game, we are likely to be the less advanced of the players in both knowl-edge and intelligence (although surely the best-looking).

Should we respond?

SETI was originally spelled CETI, for Communication with ExtraTerrestrial Intelligence. The name was changed to "Search" because searching with instruments is in the realm of science, while chatting with aliens is not. Few scientists in the field seem to have the goal of personally engaging in two-way communication, which seems sensible since signals would take many years or even lifetimes to crisscross between stars.

Whether we should respond to a signal would involve political judgments. The International Academy of Astrophysics has circulated guidelines on responding, titled *Draft Declaration of Principles Concerning the Sending of Communications to Extraterrestrial Intelligence*, which recommends letting international organizations like the United Nations decide and urges that any broadcasts we make contain content "reflecting a broad consensus." Consensus, however, is often hard to reach. It's possible that private broadcasts would be undertaken anyway—some brief ones already have been—although they might be discouraged if we discover somebody is out there.

It would be wise to learn a lot about senders before responding. It's easy to imagine dangerous information, virulent computer programs, unpleasant microbes, and other bad stuff being sent across space by malicious critters privy to our culture, technology, and genome. Fleets of invading battle cruisers don't seem very likely to appear, given that vast amounts of energy and long travel times are needed to move big things between stars, but we might not want to bet the farm against interstellar travel. One source of comfort is that we probably would not taste good to aliens, since we evolved on different worlds with different chemistries, metabolisms, and palates, although there's no money-back guarantee on this appealing logic.

It's unlikely that *all* other civilizations are nice, and some other worlds might be ruled by dark forces. Many aspects of our own civilization were imposed and are maintained by force or its threat—government, social order, tax collection, and so on—and the most aggressive individuals and groups have often ruled, sometimes brutally (see, for example, the twentieth and preceding centuries). Peaceful professors might send friendly messages but are unlikely to be the top dogs except on the most highly-evolved worlds.

Even if all civilizations were nice, we know that some individuals are not. The computer virus phenomenon is one example of how malicious individuals can wreck havoc using mere access to a communication system. If we were to communicate with other worlds, we would need to guard against either receiving or sending malicious content.

Deciding whether and how to respond would probably depend on what we received, plus a good deal of judgment.

SHOULD WE BROADCAST FIRST?

If everyone just listens and nobody broadcasts, then all searches will fail. We may decide to broadcast someday to see if we can elicit a response, or to tell our stories to whoever or whatever might be listening, or for some other reason. The idea has its own acronym, METI, for Messaging to ETI, also known as Active SETI.

We are already broadcasting unintentionally; our radio, television, and radar signals are rippling outward through a sphere of space more than 100 light years across sweeping over thousands of stars. Most of those signals would be hard to detect because they were not designed for long-distance reception, but they are out there and can't be recalled.

Humans have launched a few intentional messages. Small metal plaques were attached to the *Pioneer 10* and *Pioneer 11* spacecraft, which are speeding out of our solar system into interstellar space although thousands of years from the nearest stars. Drawings of a man and woman, the planets of the solar system, and the Sun's location are engraved on the plaques; the spacecraft themselves are messages as well. *Voyager 1*, launched in 1977, is even further out, and bears phonograph disks with greetings and music—including Chuck Berry's "Johnny B. Goode," a sure-fire crowd-pleaser. It's almost certain that these little machines will never be found by aliens; their messages are to us, that human artifacts are entering the Cosmos.

Carl Sagan and Frank Drake broadcast a three-minute radio message toward a few hundred thousand stars in the globular star cluster M13 in 1974, using Arecibo's planetary radar transmitter. It will sweep across the antennas of any listeners in that neighborhood tens of thousands of years from now, and intervening stars sooner. The message is shown in Figure 12.1.

Messages have also been sent by a commercial venture named Team Encounter. A so-called Cosmic Call was transmitted in 1999 toward several target stars on the NASA search list using a 70-meter antenna at Evpatoriya in the Ukraine—one sample page is shown in Figure 12.2.

Figure 12.1. The Arecibo message. The message was transmitted as a series of 1,679 bits by switching Arecibo's transmitter on or off. To decode it, the bits must be organized into 73 rows of 23 pixels each, which a recipient might do after noticing that 1,679 is a semiprime number, the product of the prime numbers 73 by 23. Interpreted as a crude diagram, the upside-down curve near the bottom represents the antenna, the stick figure above it shows what we look like, the ten blobs below that represent our planets and Sun, and so on. Credit: Cornell University/Arecibo Observatory.

No grand plan currently exists for broadcasting to the stars. Searchers have rarely raised the topic, possibly because it's potentially controversial and perhaps because broadcasting seems more of an engineering job than a quest for knowledge. Broadcasting to one star or a small patch of sky can be done fairly easily, but to have much chance of scoring even one listener we would

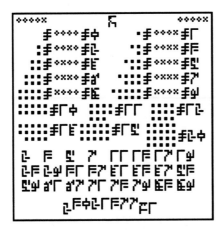

Figure 12.2. A Cosmic Call message. The number 1 is coded as a single square followed by symbols like "+++x", followed by a symbol like an upside-down L. Subsequent patterns suggest that counts of squares are being equated with binary and decimal numbers, with the symbol for equality looking like an equals sign with an S superimposed. Courtesy of Yuvan Dutil and Stephane Dumas.

probably need to transmit toward many stars for a long time and listen for responses to learn if we had succeeded.

So, *should* we broadcast first? It's hard to see much reason to, unless the effort was funded to run for a very long time and included a search to detect responses. Passive searches seem more sensible, economical, and prudent until we have looked much more carefully.

WHEN DO WE QUIT LOOKING?

It's hard to imagine any proof that intelligent life cannot possibly exist anywhere else, so as long as people are interested in exploring the universe around them they are likely to look for signals while they are at it. Future searches can always probe deeper into space for weaker signals, sift through more of the vast electromagnetic spectrum, or use smarter strategies. Even if we found nothing after searching hard, new civilizations could appear as we recently did, so it seems likely that there will always be reason to search.

Searching would not stop with the discovery of the first signal from another world because it's a good bet there would be more. With luck, we might tap into a network that tells us where to find more broadcasters. With one discovery, SETI would become a conventional science with the usual specialists and indentured graduate students laboring to make discoveries and reputations, and in time they might catalog many other worlds.

CHAPTER 13

HISTORY

MILESTONES

More than one hundred searches have sought signals from the stars since 1960. Most were very modest ventures lasting only a few hours or days, viewing few stars and listening on thin slivers of radio spectrum. A few were more ambitious, running for years and covering a large part of the sky and a bit more frequency. Yet there have been many times when nobody anywhere was listening. This chapter tours the brief history of searching, visiting what seem like major milestones and some waypoints in between.

The first milestone was the first modern search, when Frank Drake listened for radio signals from two nearby stars in 1960 with a customized shortwave receiver attached to a big radio telescope antenna. A second milestone was the first full-time sky survey, started by Bob Dixon in 1973 using the Big Ear antenna at Ohio State University with an 8-channel (later 50-channel) receiver. A third milestone came in 1985 when Paul Horowitz built an 8-*million*-channel receiver at Harvard to survey the sky for ultra-narrow radio signals. A fourth milestone was a quarter-*billion*-channel receiver at Harvard that beginning in 1995 combed a band of frequency hundreds of times wider than most prior searches. A fifth milestone was NASA's short-lived SETI program, reborn as Project Phoenix after losing federal funding, which could cover even more radio spectrum and could detect types of signals that previous searches might have missed.

Major milestones currently in sight are the Allen Telescope Array designed to search for microwave radio signals much faster than earlier efforts, and the Harvard All-Sky Search for optical pulses.

ORIGINS

The idea of listening for signals from other worlds goes back to the early days of radio. Attempts were made to find radio broadcasts from Mars in the early 1900s, with several mistaken claims of success headlined in the press. The brilliant and eccentric inventor Nicola Tesla is said to have tried broadcasting to other worlds from his secretive laboratory near Colorado Springs, Colorado, at the turn of the century, and he apparently believed that he had received signals from space as well. His high-voltage experiments created artificial lightning bolts over a hundred feet long and burned out the town's generating plant.

Searching as a scientific enterprise is an offshoot from radio astronomy, which in turn was a spin-off from whirlwind advances in radio and radar technology during World War II. Before the war, a few isolated experimenters had noticed radio noise emanating from the sky, but poor receivers limited to low frequencies prevented the pioneers from learning much about the radio heavens. Karl Jansky was the first to report radio noise coming from the Milky Way in 1932, discovering it while searching for the cause of static that hampered transatlantic radio communication when he was working for Bell Labs. Few astronomers took notice, although a young John Kraus was in the audience when Jansky presented his findings at a scientific meeting in July, 1935; Kraus would later help establish the new science of radio astronomy after the approaching war. Grote Reber, a radio engineer experimenting in his spare time, investigated the static with a homemade receiver and 30-foot backyard dish antenna in Wheaton, Illinois, beginning in 1937, and published

Figure 13.1. Left: Karl Jansky's antenna, near Holmdel, New Jersey, around 1930, with Jansky in the foreground. Image courtesy of NRAO/AUI. Right: Grote Reber and his antenna, rebuilt and displayed at Green Bank, West Virginia. Source: NRAO.

a rough radio map of the Milky Way in 1944. Those early antennas are shown in Figure 13.1.

The wartime development of radar produced much better receivers working at much higher frequencies, as well as a generation of electrical engineers and physicists eager to exploit their new radio know-how. After the war, some of them used the new insights and gizmos to build radio telescopes to investigate the unexpected radio noise from the sky and began discovering radio sources throughout our galaxy and halfway across the universe. By the late 1950s, the new branch of astronomy had developed ultra-sensitive receivers and had built titanic dish antennas—up to 250 feet in diameter at Jodrell Bank in England, an area larger than an American football field—and was revealing a universe very different than the one people were accustomed to seeing.

Big radio telescopes could also detect intelligent signals from the stars, although nobody seems to have realized it yet.

THE 21-CM HYDROGEN LINE

The radio glow of interstellar hydrogen played a lead role in the development of radio astronomy and in later searches for intelligent signals.

In 1945, a Dutch graduate student named Hendrik van de Hulst predicted that cool hydrogen atoms drifting between stars might be detectable with a radio tuned to a certain frequency, although he was doubtful that it could be done. It was known that atoms and molecules emit light when heated, and when viewed through a prism, the light displays bright spectral lines at characteristic wavelengths for different substances. The emission is caused by excited electrons falling back to lower energy levels and radiating photons whose wavelength—color, in the visible spectrum—is determined by the energy lost. Van de Hulst realized that the smallest possible change in a hydrogen atom's energy, the hyperfine transition where an electron merely flips its spin, would radiate a photon so weak that it would fall in the radio portion of the spectrum at a wavelength near 21 centimeters or 1420.405 MHz. It takes a single atom millions of years to flip spontaneously, but the vast number of them in interstellar space (roughly one in every cubic centimeter) would contribute many tiny flickers, perhaps enough to make their combined emission detectable with a radio tuned to the right frequency.

In 1951, Edward Purcell and his graduate student Harold "Doc" Ewen tested the prediction. They built a horn antenna—a roughly four-by-five-foot square funnel made of plywood and sheet metal pointed skyward—outside a window of Harvard's Lyman Laboratory of Physics and tuned a borrowed receiver near 1420. After some difficulty solved by purchasing a receiver that Ewen could customize, they detected the hiss of interstellar hydrogen on March 25. Coincidentally, van de Hulst was lecturing at Harvard at the time and so was Australian radio astronomer Frank Kerr; Ewen informed them of the discovery and their groups soon also detected the radio glow of interstellar hydrogen using Ewen's technique. All published papers simultaneously.

The discovery electrified the emerging field of radio astronomy. Soon the spiral structure of our galaxy could be seen because the galaxy *is* mainly hydrogen, and its radio glow penetrates dust that hides many of the Milky Way's stars from optical telescopes. Better yet, the velocity of the gas could be measured even in distant galaxies because the Doppler effect shifts its frequency by about five kilohertz (half the width of an AM radio station) for each kilometer per second of velocity along our line of sight. Much astronomy has been done by the radio light of hydrogen, and although many other radio lines have been found, hydrogen remains special.

MORRISON & COCCONI (1959)

In 1959, Philip Morrison and Giuseppe Cocconi published a short paper in the prestigious British journal *Nature* claiming that radio telescopes could detect radio signals over interstellar distances—signals sent by intelligent beings, not just the noise from natural sources. It was the first time such a notion had been raised in a proper scientific journal. The idea of listening for alien broadcasts was not new, but they argued that radio telescopes provided the technology needed to really do it.

The two Cornell University physicists more or less stumbled onto the idea. Morrison had been thinking about doing astronomy with gamma rays, which proved practical many years later, and Cocconi toyed with the idea that gamma rays might be used for signaling between stars. This was not entirely idle ivory tower thinking, because there was a particle accelerator in the basement of their building that could generate that kind of radiation. Good theoretical physicists, they tackled the more general question: what would be the *best* method for communication between stars?

Traveling or sending objects was rejected as too slow. Shooting charged particles like electrons and protons was rejected because their paths are bent by magnetic fields in space. Neutrinos were deemed unattractive because they are hard to detect, and other particles had other drawbacks. Light was rejected as too difficult to see next to the glare of stars. But the microwave portion of the radio spectrum seemed attractive because most stars are dark at those wavelengths, and comparatively bright microwave radio signals can be generated easily (by radars and ovens, for example). To their surprise, calculations showed that existing radio telescopes with only minor modifications could detect plausible broadcasts from nearby stars.

Radio receivers need to be tuned to the frequency of the transmitter one wants to receive, and since that was unknown, they suggested the hydrogen frequency. The idea was appealing partly because hydrogen is the simplest and most common element, and for several more practical reasons. First, all terrestrial astronomers and physicists know about the hydrogen emission line, and creatures in the same business elsewhere would likely know it, too. Second, broadcasters could improve the chances of being noticed by choosing a frequency that distant radio astronomers might observe in the course of their work. Third, receivers for that band were available at many observatories. Hydrogen was the *only* radio line known in 1959, but in 1988 Morrison still thought it was the best choice even though others had been found.

Morrison and Cocconi's paper set a number of other scientists thinking about searching for signals from the stars and gave the topic a semblance of respectability that it needed.

Morrison became an advocate for SETI, helping organize conferences, writing articles, and speaking on the subject. He died on April 22, 2005, at the age of 89, having survived childhood polio, studied under J. Robert Oppenheimer and worked at Los Alamos, witnessed the Trinity test of the first atomic bomb, and held the highest academic ranks at MIT.

PROJECT OZMA (1960)

On April 8, 1960, 29-year-old Frank Drake became the first human to listen for radio signals from other worlds with a gizmo that really could detect signals across the light years. Prior attempts in the early days of radio had little chance of success because the antennas were small and the receivers were very noisy, but Drake used an 85-foot dish antenna and one of the most sensitive receivers on Earth—a system that could detect signals as strong as some we produce, from ten or twenty light years away. He named his project Ozma, after the name of a princess in the *Wizard of Oz* series of fairy tales by L. Frank Baum.

Drake was an astronomer at the National Radio Astronomy Observatory, which was trying to put the United States on the map of the new world of radio astronomy dominated until then by Australia, England, and Holland. NRAO was building a 140-foot antenna in the hills of West Virginia near the town of Green Bank, but the project was behind schedule and it was Drake's mission to produce some scientific results quick using an existing 85-foot dish. Curiosity led him to do the calculation that showed his telescope would work as an interstellar radio receiver, and he proposed Ozma as a side project. Lloyd Berkner, the observatory head, approved it over lunch; the cost was only a few thousand dollars in specialized parts. Drake was preparing his equipment when Morrison and Cocconi's paper appeared.

A radio telescope is basically a bucket for catching electromagnetic raindrops falling from the sky, and bigger is usually better to catch as much as possible. Drake's dish, shown in Figure 13.2, was a pretty big bucket. The gentle sprinkle has to be amplified millions of times before you can hear or record it, and that's a problem because a chain of amplifiers is needed and each one adds its own noise to everything passing through—noise that gets amplified by following amplifiers, which can drown signals of interest in a sea of static. Drake managed to get a newly-invented parametric amplifier as the first and most critical link in the chain, super-quiet but notoriously temperamental. Every day he rode a lift up five stories to the focus of the antenna and tweaked the contraption until it worked right.

What made Drake's telescope an interstellar radio was a modified shortwave receiver, which allowed him to listen on a very narrow band of frequency. Most radio telescopes listen on bands millions of Hertz wide, because the noise from most natural sources hisses across a wide range of frequency and capturing more power makes the receiver more sensitive. But for com-

Figure 13.2. The 85-foot Tatel radio telescope at Green Bank, used in the first modern search for radio signals from other worlds. Source: NRAO/AUI.

munication, signals are usually concentrated in a narrow range of frequency and receivers need to be tuned to pass only that range. Drake used filters to limit the receiver's bandwidth to just 100 Hz, excluding most of the noise from the sky and receiver circuits.

He pointed his antenna at two nearby Sun-like stars—Tau Ceti and Epsilon Eridani, about a dozen light years distant—and monitored them for several hundred hours over several months, slowly tuning the receiver across the hydrogen band. He listened to the radio noise from the sky on a loud-speaker, and it traced a jagged line on a chart recorder. On the very first day he heard a strong pulsing signal that seemed to come from Epsilon Eridani, sending the recorder pens off the charts, but it disappeared after only a few minutes. The signal was detected again ten days later, and by then a second receiver with a tiny horn antenna had been pointed out the window to moni-tor local interference; the signal was just as strong in the little antenna as in the big one, a sure sign that it was local.

Project Ozma broke the ice for later searches and set a precedent for doing it at professional observatories. Drake went on to become director of the National Astronomy and Ionosphere Center at Cornell University, run-ning the 1,000-foot radio telescope at Arecibo, Puerto Rico, and he used it to map cloud-shrouded Venus with radar which was a major achievement. As a prominent astronomer, he continued to advocate SETI adding legitimacy to the field, and he later headed the SETI Institute, described later.

INTELLIGENT LIFE IN THE UNIVERSE (1966)

The Russian astrophysicist I. S. Shklovskii published a book in Moscow in 1962 titled *Vselennaia, Zhizn, Razum* (Universe, Life, Mind), speculating on the prospects of extraterrestrial intelligence. A young American astronomer named Carl Sagan added much more material for an English translation, published jointly in 1966. This influential book was packed with illustrations and facts and argued for the possible existence of life and intelligence on other worlds by drawing on knowledge from astronomy, chemistry, biology, planetary science, and many other fields. It also presented early ideas about how intelligent beings might communicate between stars.

The book caught the interest of many scientists and also reached a large general audience including your current scribbler, impressing many with its interdisciplinary overview of the evolution of stars, planets, and life—a chronicle of how things came to be, science's creation story—which Sagan later popularized in the *Cosmos* television series and books.

Sagan was a prime player in the field for over thirty years and the best known to the public, playing many important roles. He helped explain the greenhouse effect, welcomed SETI papers as editor of *Icarus: the Journal of Planetary Science* at Cornell University, co-founded the Planetary Society, convinced skeptical congressmen to support NASA's program, and more. He died on December 20, 1996, at the age of 62.

SOVIET SETI

During the 1960s and early 1970s, Russian and other Soviet Union scientists published many technical papers and books on interstellar communication, conducted a number of searches, and held several international conferences. Scientists such as L. M. Gindilis, N. S. Kardashev, V. I. Slysh, and V. S. Troitskii led searches, typically viewing only a few stars or viewing wide areas of sky at low sensitivity.

One continuing influence is a classification scheme hatched by Nikolai Kardashev, a former student of Shklovskii's, for the technological level of civilizations based on the energy they might command. A "Type I" civilization would be able to harvest all of the energy available on a single planet (sometimes taken as the solar insolation), Type II would harness the entire energy output of the star, and Type III would command the power of an entire galaxy—roughly 10^{16}, 10^{26}, and 10^{36} watts, respectively. These categories are entirely hypothetical (what would you *do* with the power of a galaxy?) and imprecise (planets, stars, and galaxies come in many different sizes), but they provide a rough scale for what users of energy might be able to do.

Terrestrial energy consumption is currently about 10^{13} watts, so we are shy of the Type I category; something like 10^{17} watts would be available if all of the solar energy falling on the surface of the Earth was collected and could be converted into usable energy.

PROJECT CYCLOPS (1971)

One day during Project Ozma, a Mooney light plane swept down through a hole in the clouds over Green Bank, West Virginia, and settled onto the observatory airstrip carrying one Bernard Oliver who was the head of research and development for the Hewlett-Packard company. Oliver had been thinking about how to search for other civilizations, had read about Drake's project in *Time* magazine, and flew in to take a look. He would become a dynamic advocate for a large-scale search for radio signals from other worlds during the next three decades—eventually heading NASA's SETI program, later helping raise it from the ashes as Project Phoenix when government funding was suddenly cut, and finally bequeathing millions of dollars to the SETI Institute upon his death.

Barney Oliver was an early recruit to Hewlett-Packard and ran the legendary electronics firm's research and development efforts for over 20 years. An electrical engineer and inventor with 60 patents, he presided over the development of the famous HP scientific calculators, high-performance computers, and sophisticated electronic test equipment. His interest in SETI and his professional stature—he was also president of the IEEE, the professional organization of electrical and electronic engineers—led NASA's John Billingham to tap him to help direct an engineering study as a first step toward a NASA program.

Billingham was the main force inside NASA lobbying for a search program. The topic was within the agency's charter, but its "life sciences" activities had been limited to keeping astronauts alive and occasionally looking for extraterrestrial microbes. Billingham fit in as a physician specializing in aviation medicine, recruited from England in 1963.

In 1971 Billingham and Oliver organized a summer workshop at Stanford University that laid the groundwork for NASA's SETI program more than ten years later and influenced many later searches. Two dozen experts in antenna design, receivers, detectors, signal processing, and other areas contributed to *A Design Study of a System for Detecting Extraterrestrial Intelligence*, the official name of their report. Oliver titled it *Project Cyclops* after the mythical one-eyed giant and for the huge circular array of big dish antennas that the report proposed.

The Cyclops report had two major thrusts. First, it identified the need for developing special detectors able to monitor millions of frequency channels at the same time and showed that such spectrum analyzers could be built using a combination of optical and electronic techniques—although within a decade microprocessors proved a better method. Second, it showed that arrays of antennas could search deep into space for intelligent signals and suggested that a program could start small and add antennas to enlarge the effort if initial searches were unsuccessful. The report concluded with an ambitious description of an array with up to 2,500 big antennas, costing some 10 billion dollars if NASA chose to go all-out, illustrated in Figure 13.3.

Figure 13.3. Artist's concept of the Cyclops array of 100-meter antennas, if it was built out from an initial smaller system. Above is an aerial view, and below is a view from ground level. The scale of the antennas can be seen from the tiny vehicle in the foreground. Source: NASA.

Over ten thousand copies of *Project Cyclops* were distributed, a bestseller as engineering studies go. It put NASA's stamp of respectability on the topic and served as a textbook for other researchers, although nearly 25 years passed before a NASA receiver went on the air.

Oliver eventually ran the NASA SETI program, and when federal funding was killed, he helped revive it as the privately-funded Project Phoenix. Billingham went on to be chief of NASA's Ames Extraterrestrial Research Division and acting chief of the Program Office for the Search for Extraterrestrial Intelligence.

THE OHIO SETI SURVEY (1973-1997)

John Kraus's Big Ear radio telescope built in the 1950s at Ohio State University was used in the first full-time search beginning in 1973 under the leadership of Bob Dixon. In addition to being the first full-time SETI program, it was also the first to survey much of the sky and was the longest running—nearly 25 years, staffed largely by volunteers with sporadic funding. Its story up to the 1977 Wow observation is related elsewhere in this book.

Ohio State sought to re-detect the Wow signal for several months in 1977 and again in 1978. Their telescope, designed for passively mapping the sky, was not a good tool for hunting a possibly intermittent source because the stationary antenna viewed the Wow locale for only a few minutes each day. Over several years, the system was modified to track objects for a few hours, moving the feed horns along tracks, but the Ohio group never returned to the Wow search. The 50-channel survey continued until 1983, after which other search strategies were tried.

From 1992 to 1997, graduate student Russ Childers scanned the radio spectrum between 1400 and 1700 MHz in a project named LOBES, for LOw Budget Eti Search. He listened on some 3,000 channels, each 100 kHz wide, hopping across 300 MHz in 20 seconds. The wide channels and short time at each frequency made the search less sensitive, but it covered parts of the radio spectrum never searched before. The sky was full of radio signals shimmering on the computer display like a radio version of the aurora borealis, many of them from satellites.

Dixon and Kraus's students, including James Bolinger and Steve Brown, developed techniques for monitoring all directions simultaneously using many small antenna elements and intensive calculations to synthesize many beams. A prototype array named Argus, for a mythological being with a hundred eyes, was designed and built by Steven Ellingson, Grant Hampson, and volunteers, producing images showing the Sun moving across the radio sky. Tuning to the right frequency and displaying all-sky images as a movie showed satellites arcing across the sky like skyrockets!

An identically named Argus project is an effort by the SETI League to organize many small amateur radio telescopes to monitor much more of the sky than is seen by big antennas.

Dixon produced many ideas that influenced the field. Invoking a principle of "anti-cryptography," he suggested that interstellar signals would be designed to be easy to identify, always doing things the simplest way. The polarization of signals, for example, would be circular because there are only two possibilities—left or right-handed—and switching between them can encode information. Many later SETI systems used two receivers, each listening on one of the circular polarizations.

Big Ear was shut down on December 15, 1997, and the antenna later demolished to build a golf course. John Kraus died on July 18, 2004, at the age of 94.

SERENDIP (1979-ON)

In 1979, a group at University of California, Berkeley led by Stuart Bowyer and later by Dan Werthimer began a search piggybacked on other astronomical observations known as SERENDIP—Search for Extraterrestrial Radio Emissions from Nearby Developed Intelligent Populations—hoping for a serendipitous discovery by looking in whatever direction astronomers selected for their own purposes. Jill Tarter, a graduate student at the time, made up the backronym to fit a word in the title *The Three Princes of Serendip*, an old Persian fairy tale which is the source of the word serendipity.

SERENDIP I was a 100-channel system installed at Berkeley's Hat Creek Observatory in 1979. Version II had 65,536 channels and ran from 1986 to 1990, mainly on the 300-foot antenna at Green Bank. Version III with 4.2

million channels ran for four years at Arecibo, beginning in 1992. SERENDIP IV, with 168 million channels, was installed at Arecibo in 1998. Version V has 128 million channels which look at seven beams and two different polarizations every ten seconds, which is somewhat (although not quite) like having 1.8 billion channels.

THE PLANETARY SOCIETY (1980)

In 1980, Carl Sagan, Bruce Murray, and Louis Friedman founded The Planetary Society, a non-profit organization to "encourage the exploration of our solar system and the search for extraterrestrial life." Sagan was a leading advocate of the possibility of extraterrestrial life and intelligence, Murray was director of the Jet Propulsion Laboratory, and Friedman had worked on deep space missions at JPL. They published the magazine *The Planetary Report,* featuring striking photographs of planets and moons sent back by spacecraft, plus popular articles by scientists working in planetary science. The society quickly grew to more than 100,000 member/subscribers.

The organization helped fund several major SETI efforts, most notably Paul Horowitz's radio and later optical searches at Harvard University and the SERENDIP/SETI@home efforts at Berkeley. It also funded many other research and exploration projects with goals such as manned exploration of Mars and demonstrating the feasibility of solar sailing (and partially supported my search for the Wow). The organization serves as a public voice for supporters of space exploration and is able to get the attention of NASA, Congress, and foreign governments.

SETI INSTITUTE (1984)

In 1984, Thomas Pierson and Jill Tarter founded the SETI Institute "to conduct scientific research and educational projects relevant to the origin, nature, prevalence, and distribution of life in the universe." Frank Drake was one of the founding directors and served as president and chairman. The institute administered $150 million in research grants during its first seventeen years, much of it contract work for the NASA SETI program, and research in the broad area of astrobiology.

One major thrust is the Center for SETI Research, initially directed by Tarter, conducting state-of-the-art searches—taking on the role that NASA once played but now privately funded. Major search projects include Project Phoenix, which conducted the cancelled NASA SETI program, and the Allen Telescope Array, described later, which aims to search hundreds of thousands of stars. Several of the institute's minor grants helped fund some of my searches for the Wow.

The other major thrust is the Carl Sagan Center for Life in the Universe, initially directed by Christopher Chyba (a former student of Sagan's), which administers NASA and National Science Foundation grants for basic research into how life began and evolved. Dozens of projects span astronomy, biology, chemistry, education, geology, planetary science, and so on—almost anything within the broad topic of life in the universe *except* searching for signals, which can be harmful or fatal to federal funding.

META II (1986, 1990)

META II was the Southern Hemisphere twin of the META search at Harvard, described earlier.

Raul Colomb and Guillermo Lemarchand of the Instituto Argentino de Radioastronomia managed to get two 30-meter antennas outside of Buenos Aires dedicated to SETI in 1986, and in 1990 equipped one with a copy of the 8-million channel META receiver. The high-resolution spectrum analyzer was used to survey stars visible from the Southern Hemisphere in the same fashion as the META survey, plus do a joint survey with META viewing the same patch of sky simultaneously visible to both sites, plus a targeted search of 71 nearby stars. Colomb was the director of the institute; Lemarchand was his student who became the SETI program director.

In 1997, the system was upgraded to cover a wider range of frequency (2 MHz, up from the original 0.4 MHz) and to cover three different bands at 1.4, 1.6, and 3.3 GHz. In 2007, an upgrade to a 256-million channel SERENDIP V spectrum analyzer was planned but delayed due to lack of funds.

NASA'S SETI PROGRAM (1992)

NASA's search was launched on Columbus Day, 1992—the 500th anniversary Christopher Columbus's arrival in the New World—but it was not destined to stay afloat long.

It took a decade to get on the air because NASA's approach was to develop a state-of-the-art interstellar radio receiver system using custom microprocessor chips, advanced signal detection techniques, interference databases, and more. Initially known as the Microwave Observing Project (MOP), it was later called the High Resolution Microwave Survey (HRMS); both names oddly neutral and uninformative, neither acronym catchy.

John Billingham and Barney Oliver led the project, along with Jill Cornell Tarter. With a doctorate in astronomy from University of California at Berkeley, Tarter had been involved with SETI as a graduate student and later as a post-doc at the Ames Research Center where NASA's program had been gestating for many years. The program was to consist of a "bi-modal search strategy" in the inevitable agency jargon; "two-part" to the rest of us.

One part was a Targeted Search to examine 750 fairly nearby stars using the biggest and most sensitive antennas available: the 1,000-foot at Arecibo, the 210-foot at Parkes, Australia, for the Southern Hemisphere, and others. The search was to cover 1 to 3 GHz in chunks of ten million 1-Hz channels, listening for up to a thousand seconds on each frequency band and each star, and analyzing the torrent of data in real time looking for signals of interest. The target list included stars within about 150 light years and of the same general age and spectral type as the Sun, plus all stars within 25 light years—just in case Sun-like stars are not the best sites for life after all. The targeted search was under the direction of NASA's Ames Research Center.

The second part was an All Sky Survey to scan the sky over a wider range of frequency from 1 to 10 GHz. It was to cover most of the sky, although a thousand times less sensitive than the targeted search because it would not listen long in any one direction and would use smaller antennas. It would have taken about seven years, scanning the sky some thirty times with receivers optimized for various frequency bands, able to detect a strong enough beacon anywhere in the northern skies—not just selected nearby stars. The sky survey was under the direction of the Jet Propulsion Laboratory and was designed to use big antennas at NASA's Deep Space tracking facility at Goldstone, California.

Both searches were sensitive enough to detect broadcasts intended for an interstellar audience and perhaps strong local signals like radar. The sky survey could detect signals like those transmitted by the Arecibo telescope's planetary radar out to 100 light years, and the targeted search could detect them at over 1,000 light years.

NASA's search was abruptly terminated in a 1993 congressional committee action. Senator Richard Bryan (D-Nevada) attacked the project as "foolish and wasteful," and NASA was effectively forbidden that line of research. The $12 million annual program had barely begun searching, but it was time for the teams at Ames and JPL to look for other jobs and abandon the $60 million investment in hardware and software they had spent a decade building.

PROJECT PHOENIX (1994)

NASA's project was partially reborn within a year with private funding as Project Phoenix, named after the mythical long-lived bird reborn from its own ashes every 500 years. The SETI Institute hired many of the original workers and got the NASA equipment on loan.

Barney Oliver was the magician who conjured up the $7 million needed quick. He was acquainted with many founders of the electronic and software industries, and they responded to his solicitation by giving roughly $1 million each toward a first-year budget: Gordon Moore, co-founder and chairman of Intel; Paul Allen, co-founder of Microsoft; both William Hewlett and David Packard; Mitchell Kapor, founder of Lotus; and Oliver himself.

Project Phoenix continued the targeted search planned by NASA, but the sky survey was dropped because it needed NASA's forbidden antennas.

Phoenix's receivers were housed in a trailer that was shipped to various radio observatories and connected to the big antennas for weeks or months at a time. The Phoenix team worked through a list of target stars, tracking each one for several hours, stepping from one band of frequency to another every few hundred seconds. Signals appeared often and the computer looked them up in a database of known interference and ignored them if they appeared there. The system was periodically tested by pointing the antenna toward the *Pioneer 10* spacecraft at the edge of the solar system—the most distant radio signal available—and detecting that weak signal proved the system was working, as shown in Figure 13.4.

When a promising signal appeared, a second smaller telescope (inelegantly dubbed the FUDD, for Follow Up Detection Device) at a distant site tuned in to help rule out local interference. Additional tests, like pointing away and then back, further screened out interference.

In 1995, Phoenix observed from the 210-foot Parkes radio telescope in New South Wales, Australia, viewing stars visible only from the Southern Hemisphere. From 1996 to 1998, it encamped at Green Bank, using the aging 140-foot telescope half-time as conventional radio astronomy operations

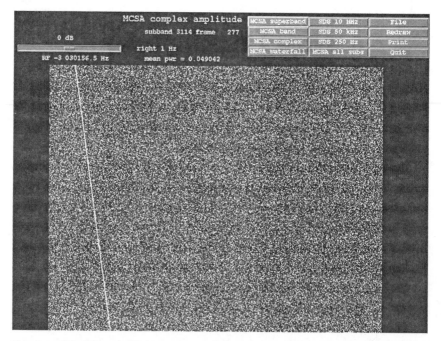

Figure 13.4. Signal from the *Pioneer 10* spacecraft traveling near the edge of the solar system. The display shows a rising stack of successive spectra recorded at about one-second intervals, each a horizontal line of dots representing 924 channels spanning 643 Hz, with the signal strength in each channel indicated by brightness. *Pioneer's* radio signal is the diagonal line, drifting in frequency because of Doppler effects due to the Earth's motion. The transmitter's power was just a few watts (comparable to a flashlight) from over 10 billion kilometers (10 light hours) away. Courtesy of the SETI Institute.

wound down pending a move to the new 100-meter Green Bank Telescope. In 1998 and several later years, Phoenix observed for roughly a month annually using the Arecibo antenna.

When the Phoenix project was completed in 2004, it had listened for more than 11,000 hours to more than 800 stars with some of the biggest antennas on Earth, covering over a billion 1-Hz-wide channels from 1.2 to 3 GHz with enough sensitivity to detect both intentional broadcasts and some unintentional signals like radar. Its ambitious successor was the Allen Telescope Array, described later.

Oliver died on Thanksgiving evening 1995 at the age of 79, leaving $20 million to the SETI Institute, intended to help support a long-term search without the uncertainty of government funding.

BETA (1995)

The META project (Megachannel ExtraTerrestrial Assay) run by Paul Horowitz at Harvard from 1985 through 1994 is described in another chapter.

In October 1995, Horowitz began a more ambitious search, naming it BETA for its (very roughly) *billion* channel spectrum analyzer. The 240-million channel machine was designed to cover the entire water hole from 1.4 to 1.7 GHz, hopping through the spectrum in eight 40-MHz chunks, listening for two seconds at each. Taking only 16 seconds to scan the band, it could make eight scans during the two minutes a source drifted through the beam, potentially detecting a signal repeatedly to see if its strength matched the antenna pattern.

Borrowing Ohio State's dual-beam approach, the antenna was equipped with two feeds to create two side-by-side beams. Celestial sources drifted through one beam, then through the other beam a few minutes later. If an interesting signal seen in the first beam reappeared in the second beam, the antenna would begin following the source, pointing away and returning repeatedly. A good candidate would repeatedly show up in the beam moving like the stars, but would fade when the antenna pointed away.

In March, 1999, the antenna was damaged by high winds, and it was dismantled in 2007.

SEVENDIP (1997-ON)

Two optical searches have been mounted from the University of California at Berkeley, under the name SEVENDIP or Search for Extraterrestrial *Visible Emissions from Nearby Developed Intelligent Populations*.

One search, directed by Geoff Marcy, reviews data from target stars gathered by planet searches, looking for spectral lines that might be due to lasers shining our way. A second search, directed by Dan Werthimer, looks for billionth-second pulses of light from several thousand nearby stars using a 30-inch telescope—another way in which interstellar communication might be done using lasers.

HARVARD OPTICAL SETI (1998, 2006)

Paul Horowitz and associates at Harvard have mounted optical searches, consisting of a Targeted and an All-Sky search. Both exploit ultra-fast electronics to search for nanosecond-timescale pulses of light, assuming that broadcasters use billionth-second flashes of light to briefly outshine their home stars.

Horowitz began a targeted optical search in 1998, riding piggyback on a planet search conducted by David Latham and Robert Stefanik using the 61-inch telescope at the Harvard-Smithsonian Oak Ridge Observatory and observing some 5,000 stars by the end of 2000.

The all-sky search began in April 2006 using a 1.8-meter (72-inch) mirror to do a transit survey, viewing far more stars than a targeted search. Using a spherical reflector, the light from the stars drifting overhead converges along a blurry focal line that sweeps over two arrays of 512 photodetectors counting photons a billion times a second. Stars over a 1.6° x 0.2° field are viewed for roughly a minute, and the system looks for brief spikes in the intensity of light that might indicate a laser or other optical signal pointed our way; two arrays are used to make sure spikes come from the sky rather than from a single twitchy detector or cosmic ray. The system harvests terabits of data every second—equivalent to all books in print—before the obvious chaff is sifted out.

SETI@HOME (1999)

In 1999, a venture billed as SETI@home caught the public eye and interest. It sent small chunks of data from the SERENDIP search at Arecibo to volunteer computers via the Internet to do much more extensive calculations than were otherwise possible. To volunteers, the program looked like a screen saver using a computer's resources when the owner did not, as in the example shown in Figure 13.5. During the first month, over a half-million people participated in the project, and within a few years, millions had joined, doing a thousand years of computing every day. It was the largest distributed computing project ever undertaken until that time, mustering more computing power than the biggest supercomputer. Free.

The scheme was thought up by computer scientists David Anderson and David Gedye in 1994, working with astronomers Dan Werthimer, Woodruff Sullivan, and eventually others, to crunch the numbers produced by the SERENDIP project described earlier.

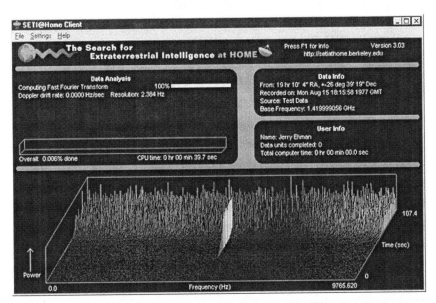

Figure 13.5. A sample SETI@home screen, showing noise power spikes rising up from many narrow channels over time. The foreground shows an example of what a strong signal might look like (a simulated Wow signal taking 12 seconds to drift through Arecibo's beam), while the background shows the usual noise (lower in the foreground because the vertical scale is compressed to accommodate the strong signal). Courtesy of Bruno Moretti Turri.

SETI@home analyzes 2.5 MHz around the hydrogen frequency, and the screensaver program processes small chunks of it looking for signals with many different channel widths (0.075 to 1220 Hz) and many different frequency drift rates (-50 to +50 Hz/sec). The goal is to hit upon the optimum receiver settings, such as a bandwidth matched to the signal to get good sensitivity, and a drift rate that keeps a signal in one channel long enough to add up to a detectable amount of power. The scheme increases the chances of detecting signals with unknown characteristics, but requires lots of computing. Results are returned to a central computer and researchers later screen them for anything of interest, especially looking for the Gaussian rise and fall of intensity expected from objects passing through the antenna's beam. Some promising candidate positions have been re-observed, but so far none have panned out. The search may eventually cover the roughly 25% of the sky visible to the Arecibo antenna.

In 2008, a new search named Astropulse was integrated with SETI@home, to search for very brief pulses that narrowband searches would be unlikely to find—pulses lasting for only one-millionth of a second, rather than for seconds or longer.

ALLEN TELESCOPE ARRAY (2000, 2007)

In 1999, the SETI Institute announced plans for a One Hectare Telescope nicknamed "1HT," a large number of small dishes designed to have a collecting area of 10,000 square meters (the hectare is a metric unit of area, about 2½ acres, larger than a soccer or football field). That's as big as the world's largest fully steerable radio telescopes, although smaller than the partly-steerable Arecibo antenna with 73,000 square meters. It was designed to search many more stars than Phoenix and much more quickly.

In August 2000, the project was renamed the Allen Telescope Array, initially funded by Microsoft co-founder Paul Allen's donation of $11.5 million for research and development, and former Microsoft Chief Technology Officer Nathan Myhrvold's donation of $1 million for the electronics laboratory. The array is being built at the University of California Hat Creek Observatory several hundred miles northeast of San Francisco and is designed to provide

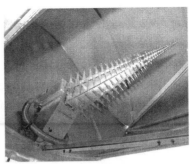

Figure 13.6. A group of ATA antennas (above left) and a shrouded feed (right), which contains a cooled low noise amplifier. Photos courtesy of Seth Shostak.

Figure 13.7. Depth into space of planned ATA searches, compared with the depth of the previous Project Phoenix. Illustration courtesy of the SETI Institute.

radio imaging like the Very Large Array and to be easily expandable—rather reminiscent of Cyclops but on a more modest scale.

The design calls for hundreds of 6.1 meter (20-foot) dishes, much smaller than the giants of conventional radio astronomy and cheaper because they can be mass-produced. Linked together as a phased array, the antennas can all point toward one target area for maximum sensitivity, or point in different directions to view several different targets at lower sensitivity. Part of the system is shown in Figure 13.6; 42 antennas have been built in a first phase of construction and function as a partial system.

The ATA aims to search hundreds of thousands of target stars reaching deeper into space, rather than the roughly thousand stars during Project Phoenix, as shown in Figure 13.7. With fairly small dishes, the array will have a wide field-of-view (2.5° at 21 centimeters, five Moons across) and will use digital processing to form multiple pinpoint beams to listen to more than one star at a time—while at the same time allowing conventional radio astronomers to look at areas of interest to them. The system will be able to search 0.5-11 GHz, covering the entire terrestrial microwave window and examining 100-million-channel chunks of spectrum. More channels can be added as the price of computing falls, and more antennas can be added to search deeper in space.

In April of 2011, operations at Hat Creek were suspended due to state of California and Berkeley budget cutbacks. Funds for temporary operation were raised through a public appeal and longer-term support was being sought, but the future of the ATA was uncertain without $1 or 2 million annually for operations. The first 42 antennas cost about $50 million which was largely research and development; building the system out to 350 antennas for world-class performance would require another $50 million.

SETI 2020 (2002)

In 2002, the SETI Institute published *SETI 2020*, a thick book looking down the road to assess the prospects for tomorrow's equipment and strategies for finding extraterrestrial signals—a broad update to the original *Project Cyclops* report. Dozens of experts contributed, covering many topics in lots of technical detail.

The main message was that searches in both radio and optical realms are appropriate using current techniques, and that new technologies should be developed that are known to be better but not yet ready for prime time. The SETI Institute itself was funding major radio and optical searches, as well as some of the future-oriented research they advocated.

SUMMARY

SETI has come a long way since 1960—from one channel to hundreds of million, from searching a few stars to all-sky surveys, from a few part-time professors to Programs. It has not discovered anything yet, but it has developed instruments and techniques that could one day detect a distant signal and answer YES! to the question "Is there intelligent life on other worlds?"

No radio search has yet looked at the entire microwave window in any direction, much less over the entire sky, and optical surveys are just beginning. No search has examined more than a thousand stars for very long with high sensitivity, and no search has been sensitive enough to detect local radio communications of another world. And very few searches have dwelled long enough to detect signals that appear intermittently—for example, once each "day" as a source planet rotates, sweeping a beam across the sky like a lighthouse.

The twenty-first century opened with big steps forward in both radio and optical searches, although at the million-dollar scale rather than the billion-dollar scale that has preceded major discoveries in areas like astronomy and physics. Most current searches are privately funded, which may be appropriate for a science still in search of a subject, but relatively modest financing limits the pace of progress.

CHAPTER 14

WORRIES

Searching for life and intelligence on other worlds is fraught with worries; people have thought up lots of possible reasons why life might not exist elsewhere, or might not be intelligent, or might not broadcast, or might not want to be found. This chapter presents pessimistic views that make some folks think searching for intelligent life elsewhere—or even life of any kind—is unlikely to succeed.

Many worries arise because we have barely begun exploring other planets, and because we don't know how much of our own story depends on details of our planet's particular history. As we learn more about other worlds and more about our own, we should become better able to judge the prospects for life and intelligence elsewhere.

NO EXTRASOLAR PLANETS

Extrasolar planets are a good example of a worry with a happy ending. Until the 1990s, the existence of planets around other stars was a big worry because nobody had been able to find any despite numerous attempts. Some people were pessimistic because of the failure to find extrasolar planets, while others were optimistic because theories of star formation suggested that planets would form at the same time as stars. We now know that extrasolar planets do exist, discovered by searches using improved instruments and techniques, and the sky seems to be *full* of them.

NO OTHER SYSTEMS

Judging from the first planets discovered around other stars, it appeared that planetary systems like ours might be rare. The planets we found were usually bigger than Jupiter, and only one was detected at each star, quite unlike our solar system.

The seeming predominance of single giant planets was a consequence of early search techniques that were most sensitive to massive objects, unable to "see" Earth-size planets at all. Planets the size of the Earth were detected as methods improved, and multiple-planet systems were found—four circling the star Upsilon Andromedae, five circling 55 Cancri, and more elsewhere— so these worries are no longer ominous. Improving instruments and techniques are letting us detect more smaller planets and multiple-planet systems, although it may be some time before we know if our system is typical or somehow unusual.

UNSTABLE ORBITS

Simulations of planetary system orbits show that some are not stable; the gravity of giant planets like Jupiter can scatter smaller planets into distant, frigid orbits or out of a system entirely. In our case, the giants are docile and helped sweep up debris that otherwise might bombard us, but it's unknown how often things play out that way.

Another worry is binary stars because one-third or more of all stars have a companion, and it's known that in some configurations planets would be thrown out of multiple-star systems. But some multiple star configurations are stable, such as when planets are fairly far away from stars that are close together, or when the stars are far away from each other and planets orbit one of them relatively close in.

BIG MOON NEEDED

Our Moon is large compared with the size of the Earth—more than a quarter of its diameter, although 1/81 of its mass because it's less dense—and its influence is so pervasive that some think life might not exist here without it. We see other big moons in the solar system, but circling much bigger gas giant planets; the more nearly Earth-scale planets Mercury, Venus, and Mars have no significant moons (Mars has two tiny ones). Our moon is thought to have been formed in a collision with something the size of Mars, which is presumably a rare event, so such big moons might be rare, and likewise life.

One major effect of our Moon is to help stabilize the tilt of the Earth's axis. The 23.5 degree inclination of the axis keeps the seasons relatively mild; without a big moon, the axis might tip further over millions of years, making seasons much more severe for much of the globe. Another effect is tides, which were much higher in earlier times when the Moon was much closer, and the tides would have washed seawater over large costal areas, perhaps stirring and spicing primordial soup as life was emerging.

This worry is hard to dispel but does not seem fatal. The extent of the Moon's role in stabilizing the Earth's axis and climate is not known, and it's possible that relatively big moons are not as rare elsewhere as they are in our solar system.

SCARCITY OF BUILDING BLOCKS

Life on Earth depends on a large number of elements and compounds, many of which might not be available on all planets, or not in sufficient quantity on the surface. Planets that formed early in the universe, for example, are probably short of the heavier elements needed for our style of life, as are gas giants like Jupiter and Saturn.

It could turn out that the Earth is unusual in having a lot of carbon, for example, which is needed to form the organic compounds required by life as we know it. But we don't have strong reasons to think we are very special in our carbon endowment because Venus and Mars also have it (although the Moon is short of it). We see organic compounds on the Earth and Titan, but not on Venus and Mars, perhaps because they have been destroyed by heat or chemistry. This is a worry about how often life could appear rather than whether it can appear anywhere else, but if carbon is rare elsewhere, so might be life as we know it.

IMPROBABILITY OF SELF-REPLICATION

The odds of self-replicating molecules emerging by chance alone seem very small, and it has been shown that molecules can't get very big and complex floating on the surface of a primordial pond because they get broken apart by ultraviolet light.

This is not too disturbing because many other factors were probably at work on the early Earth, such as shade from ultraviolet light, catalysts, and deep-sea volcanic vents, for some examples. The origin of life on Earth is lost in the vast reaches of time and erosion, and although much research is focused on early times, it is possible that we will never be certain about how life began. The fact that we don't yet understand how life emerged here means that we can't be certain that it happens elsewhere, but that does not seem like a good reason to not look for life elsewhere.

COMPLEXITY OF ANIMALS

Some scientists think that single-cell life might be common on other worlds but that bigger and more complex life, like animals, might be rare—a view sometimes known as *Rare Earth,* after a book by Peter Ward and Donald Brownlee. The fact that life on Earth was single-celled for 3.5 billion out of 4 billion years is a very good reason for this worry. We don't know what caused the Cambrian explosion of multi-cellular organisms some 530-540 million years ago, although an earlier period of ice covering much or all of the planet may have played a role.

We don't know of anything that would prevent the evolution of complex life on other planets, although there is no reason to think it will always happen. If simple life begets complex life only rarely or only after a very long time, then we might need to look at older stars or look deeper in space at more stars, but it doesn't seem to be a show stopper.

RADIATION

Many kinds of radiation could have prevented life from getting started or could have killed it off at any time—perhaps we've been lucky to avoid being zapped, and perhaps others have not been so fortunate.

Ultraviolet radiation from the Sun would be fatal to much life on the surface of the Earth if it was not sheltered by a fairly thick atmosphere, and planets with less protection might be sterilized by UV. Similarly, cosmic rays can shatter genes that aren't protected by oceans of gas or water overhead.

Gamma ray bursts from nearby supernovae are a worry because the surface of the Earth could have been sterilized or the atmosphere damaged enough to let in much more radiation at any time in its history and perhaps many times, and life elsewhere would face the same risk. It's been suggested by astrophysicist James Annis that potentially lethal gamma ray bursts were much more frequent earlier in the history of the galaxy—perhaps every few million years, rather than every few hundred million years now—and might have fried life before it evolved very far.

Genesis deep underwater is one possibility that would shield emerging life from radiation and might allow life to get started on planets and moons without the protection of a thick atmosphere.

GALACTIC HABITABLE ZONE

It's possible that not all parts of our galaxy are conducive to life, so that not as many other civilizations exist as simple estimates suggest.

A number of authors including Charles Lineweaver and Bela Balazs have argued that conditions suitable for life may be limited to a fairly narrow ring around the center of our galaxy where we happen to reside—where enough heavy elements have been created by supernovae to form rocky planets, but supernovae are not as frequent as in the more crowded core or spiral arms where life might be fried before it evolved long enough to get smart.

The existence of such a galactic habitable zone would reduce estimates of the total number of civilizations in the galaxy, but would not have much impact on the rationale for current searches because the zone is thought to be thousands of light years wide.

PLATE TECTONICS

The Earth is alive, geologically speaking, with a molten core and plates of crust constantly sinking back into the mantle at their edges—carrying away carbon in rock later to be recycled by volcanoes as carbon dioxide. Without that greenhouse gas, the planet might freeze, and we don't see signs of current plate tectonics on Mars or Venus.

This seems like a serious worry, because greenhouse gas is though to be needed to keep temperatures suitable for life over billions of years. Our molten core is due to both size of the planet (smaller ones like Mars cool much faster), and radioactive elements like uranium that help keep things toasty (but aren't as abundant everywhere in the galaxy).

FERMI PARADOX

Enrico Fermi, designer of the first self-sustaining nuclear fission reactor in 1942, asked colleagues over lunch one day in 1950, "Where is everybody?" His point seems to have been that *not* having seen alien visitors might be significant—it might tell us something about the feasibility of interstellar travel, or as some have interpreted the story, it might tell us whether extraterrestrial civilizations exist at all. Two of three people present (Edward Teller and Herbert York) recalled many years later that the thrust of conversation was the improbability of interstellar travel, but Fermi's question has become linked with the idea that the absence of aliens everywhere is evidence that they do not exist anywhere in our galaxy—the so-called Fermi Paradox.

Astronomer Michael Hart and physicist Frank Tipler have developed the idea that a sufficiently advanced civilization could colonize the entire galaxy in a few million years, which is a small fraction of the billions of years during which the galaxy has been around, and since we don't see them here in our solar system, such civilizations must not exist.

This argument is worrisome to some people because it might be *possible* for spacecraft to travel between stars, and *possible* for colonies to send out more colonizing ships (maybe self-replicating robot machines), and *possible* for this process to fill the galaxy fairly quickly.

One weakness in the scheme is that it depends on unobtainium—engineer's slang for materials or devices that would be perfect for some purpose, except that they do not exist or are prohibitively expensive. But none of this seem forbidden by physics, so the fact that we can't do such things does not disarm the argument. Another weakness is assuming a motivation to colonize other worlds across the vast distance between stars, and that motivation might not exist anywhere. But humans have spread across the Earth and populations can expand exponentially, so it's hard to argue that others who are more advanced might not similarly spread across the stars—although it would be a vastly more complex and expensive undertaking.

The fact that we don't see extraterrestrials everywhere seems like a weak reason to conclude that they don't exist anywhere. They might exist but not be able to fill the galaxy, or might not want to, or might have already filled it with artifacts we have not yet found or recognized. It's possible that nobody *ever* spins off colonies or swarms of self-replicating robots into space, since doing so would create the possibility of conflict with the home world—a risk that intelligence might avoid.

UNIQUENESS OF OUR INTELLIGENCE

Virtually all of the Earth's history is evidence against intelligence like ours evolving. It appeared in the last geologic instant and in only one species;

everyone else seems to get along without the fancy thinking organ we have between our ears.

We know of no reason why minds something like ours can't emerge in other species, but we seem to be unique—especially in our complex language and abstract thought—and the idea of other minds like ours is hypothetical. This will be a worry until we find evidence of Others, but we can take some comfort in the example of Neanderthal and wonder if animals like chimpanzees have the potential for traveling the same road.

ALL OF THE ABOVE

Physicist Steven Webb wrote a popular book titled *Where Is Everybody?* considering fifty possible solutions to the so-called Fermi paradox, and concluded that our Earth is the only home in our galaxy to folks with whom we might have anything to talk about.

His guess (as he called it, which is the best anyone can claim) was that many of the worries cited here act to "sift out" many of the cases where life does arise, and sift out many of the cases where it does evolve to some complexity, so that not many worlds are left with creatures to develop language and its consequences, like culture and technology.

It seems likely that some or many stories of life arising and evolving have tragic endings, but we don't know enough to assign probabilities to various events and conclude that we are the only story with a happy ending.

EVOLUTION

Intelligence on some other worlds may have been around much longer than we have been, and even if it was like us at one time, evolution might have produced very different minds or machines or something else we can't imagine.

Evolution carried us from monkey-like to modern in a few million years through tens of thousands of generations, and another million years might produce a similarly big change. Cultural evolution moves much faster; today's knowledge and technology is vastly different than one hundred years ago, and may change even more in the next century. Others elsewhere might be at very different stages than we are—stages that don't include an interest in communication or doing anything that we could detect from a distance.

This sort of worry reminds us anything we find will almost surely be surprising, but is no reason for not looking.

FAILURE TO COMMUNICATE

It's possible that we will find minds something like ours, but that differences in senses or other stuff will make understanding difficult or impossible. One reason for pessimism is that attempts to communicate with seemingly intelligent species on Earth, such as dolphins and chimpanzees, have not been very successful.

What sort of "language" would creatures develop if their primary senses were based on smell (like much of our animal world), bioluminescence (like fireflies), electric fields (eels), echolocation (bats), magnetic fields (some birds), or senses we can't imagine? How would we deal with a message that specified a sequence of pheromone chemicals intended to be... *sniffed*? Some extraterrestrial minds might be profoundly different from ours and might prove utterly inscrutable, but there is no reason to expect that in all cases.

PROGRESS

It is possible that there are better ways to signal between stars than we currently know about or can manage—ways already known and used by the folks we'd like to find.

The basic physics that makes the radio and optical spectrum look attractive seems unlikely to change dramatically, although that's what people always think, isn't it? An often-cited hundred-year-old quote attributed to the physicist William Thomson (Lord Kelvin) is: "There is nothing new to be discovered in physics now. All that remains is more and more precise measurement." He was proven spectacularly wrong by the 20th century that followed his pronouncement.

The technology we use for searching is constantly improving, and it's possible that broadcasters assume that searchers use technology that's better than our current best. One example is that we can already monitor the entire sky continuously with low sensitivity, but sometime in the future we will probably be able to do it much better. If all broadcasters assume that searchers use that future technology, they might send signals very rarely—say, once every few years—and all of our current searches would fail.

ECONOMICS

Why would anyone go to the expense and trouble of broadcasting, when they might receive little or nothing in return? How might the cost of broadcasting affect the kinds of signals we can expect?

Trade in information is one possible motivation, and we can imagine others. It's possible that energy and big antenna systems are cheap on some other worlds, but in most places the cost of power and antennas probably constrains broadcasts. It's possible that we have gotten the relative scales of broadcasting and searching wrong, and that searchers, who want information and may benefit from it, must mount larger-scale efforts. Barney Oliver argued that the cost of broadcast and receiver systems would be equal in an optimum system, and Gregory, James, and Dominic Benford have argued that cost-conscious beacons might produce infrequent pulses, which would require searchers to listen longer than is the usual practice.

We have seen that the power requirements for some kinds of broadcasts are very high—especially beacons that illuminate all directions all of the time—so that high cost (or impossibility) might force some or many broadcasters to use other strategies such as intermittent signals.

ZOO

Some have suggested that aliens may exist and be aware of us, even able to signal or visit us, but do not because there's a "Don't Bother The Animals" sign posted.

There is no evidence of such a Galactic Society making itself hard to find, but if we thought there was one, it would be a strong reason to search for it. They might not shout at us with beacons, but we might detect their local radio use, big engineering projects, life-gases in their atmospheres, or other signs of their presence.

BLASÉ

Maybe they are out there and aware of us and able to communicate but *just aren't interested.*

That could be true in some cases, but there is no reason to think they would all be of the same mind; some might be chatty or curious, even if some are not. It is conceivable, though, that intelligence passes through our stage and in hypothetical later stages has no interest in communication—at least not with the likes of us. If there are no beacons out there, it would make finding anyone harder, but we might detect the presence of taciturn critters by methods that don't require them to be attempting communication, as in the Zoo scenario.

CABLE

It's easy to imagine a future era when the Earth will be a quieter place, radio-wise—when radio and television broadcasting are replaced by wired, satellite, or cellular systems that would be very difficult or impossible to detect from other stars. More advanced worlds might leak less radio energy and be harder to find.

This does not seem like a big problem today because our searches usually seek broadcasts intended to span the stars; they don't aim to detect local chatter. But it may be a worry in the future if we fail to find intentional broadcasts and need to search for unintentional signs of technology. Even then, eavesdropping on local communication would not likely be our goal, since it probably would be easier to find other stuff like radar, which must go out "over the air."

VASTNESS OF SPACE

The immense distance between stars is a big worry even for folks who think broadcasts are out there. Not only are distant signals generally harder to detect because they weaken with distance, but more distant broadcasters have many nearer stars than ours to choose as targets.

Our galaxy is roughly 100,000 light years across and 1,000 thick, bulging toward the center. If our searches reach out 1,000 light years in every direction, then we cover less than one-thousandth of those stars, and over a thousand broadcasters would have to exist scattered across the galaxy in order for us to have a good chance of having one within range. There could be lots of chatty creatures out there, but none near enough to hear.

This worry is eased by the large number of relatively nearby stars—over a million within a thousand light years—and by the fact that some searches could detect strong enough signals from virtually any distance.

IGNORANCE

We don't know a lot about the universe yet ("it's a mighty big place!") and our limited knowledge and experience are handicaps. Are there types of life or intelligence that we've failed to foresee? Places other than planets orbiting stars where folks might dwell? Messages rippling by or even lying around on the ground that we fail to recognize? Are we *them*?

It's hard to know what we don't know, except that it's a lot, but pleasantly we learn more as time passes. Almost every new day brings news of discoveries in astronomy, biology, and other areas that add to our store of knowledge and understanding; the general trend suggests that we do have a pretty good bead on things but should expect surprises.

SETI SEARCHES FOR GOD

Searchers can be said to be looking for powerful beings in the sky, and some might worry that searchers are guided by unconscious notions about deities arising from religion. But most people searching for ET are astronomers whose job is to search for things in the Heavens—things including planets as well as stars, galaxies, and much more—so it seems reasonable that they would look for biology, too.

It's fun to turn this around and ask if an alien searching for radio signals from other stars should be suspected of looking for their version of God, assuming they have one. We know that their search would be logically sound because they might find us with enough effort, although we would probably prove disappointing in the deity department.

Summary

Worries that make the existence of life and intelligence elsewhere seem dubious may turn out to be justified in some cases, but the cosmic dice have surely rolled in many different ways—possibly yielding smart critters in some other places, aware of the universe beyond their immediate grazing or hunting ground, and motivated to learn more about it. They and their worlds would surely be different, but perhaps sufficiently similar to us in some cases to bring about contact.

The sensible thing seems to be to keep looking to find out what's out there and perhaps discover surprises.

SETI EPILOGUE

The first time I went snorkeling along the Florida Keys, I lost my car keys in the Atlantic Ocean. I had been gliding through warm water for hours marveling at the seascape—following fish, fanning sand, uncovering live conches, exploring an alien world just a few feet below the surface of our own. As the Sun set, my family and I returned to our shoreline campsite, and I realized that the keys had fallen out of my swimsuit pocket and were somewhere out in that big ocean. It was getting dark and cooling off fast. Our clothes were locked in a rented car, we were a hundred miles from a spare key, and a thousand miles from home. The children were shivering and my wife was frosty, so I waded back into the Atlantic and found the keys underwater half-covered in shifting sand.

On another occasion, my wife realized an earring was missing after a long walk through LaFortune Park in Tulsa, Oklahoma; she retraced her steps and found it buried in gravel in a big playground where our children had played.

Finding lost things is not the same as finding ET, but using strategies like the ones people use to find things may help us discover signals from the stars if there are any. Logic, intuition, and determination—plus luck—can beat long odds, although the lack of success so far shows that it's no easy task.

I think it's fair to say that we've barely begun searching seriously, because only small segments of the spectrum have been examined and in most directions for only a matter of minutes. Our present level of effort amounts to roughly a dozen scientists and a few million dollars per year—one-thousandth of what it took to reach the Moon. The astronomer and SETI scientist Jill Tarter put it this way:

> The amount of searching that we've done in 50 years is equivalent to scooping one 8-ounce glass out of the Earth's ocean, looking and seeing if you caught a fish. No, no fish in that glass? Well, I don't think you're going to conclude that there are no fish in the ocean. You just haven't searched very well yet.

If we remain empty-handed after a lot more effort, including searches of the spectrum over extended time periods looking for intermittent signals, then a more pessimistic view of the prospects for finding interstellar broadcasts will be in order.

Failing to find signals would not prove that life or intelligence does not exist elsewhere, because SETI can only find gizmo-builders generating signals. Other kinds of searches may find silent signs of extraterrestrial life, such as fossils on Mars or life gases in the atmospheres of planets circling other stars. But failing to find other voices out there would be disappointing, since succeeding would be so interesting!

Bibliography

Chapters 1 & 2. The "Wow" Signal and A Candidate Interstellar Signal

Ball, J. A. 1976. In *Methods of Experimental Physics*, Vol. 12 Astrophysics—Part C: Radio Observations, Meeks, M. L. (ed.), Academic Press, 308–314.

Dixon, R. S. SETI in the 1970s, web page last modified August 13, 2005, http://www.bigear.org/oldseti.htm

Dixon, R. S. 1985. The Ohio SETI Program—The First Decade, *in IAU Symposium 112—Proceedings*, Reidel, Dordrecht, Holland.

Dixon, R. S., and Cole, D. M. 1977. A Modest All-Sky Search for Narrowband Radio Radiation Near the 21-cm Hydrogen Line, *Icarus* 3, 267–273.

Dixon, R. S. 1973. A Search Strategy for Finding Extraterrestrial Radio Beacons, *Icarus* 20, 187–199.

Dixon, R. S. 1970. A Master List of Radio Sources, *ApJ Supplement Series* 2 Nr. 8, and subsequent updates.

Dixon, R. S., and Kraus, J. D. 1968. A High-Sensitivity Survey at North Declinations Between 19° and 37°, *AJ* 73 No. 6.

Ehman, J. 2011. Wow! - A Tantalizing Candidate, in *Searching for Extraterrestrial Intelligence: Past, Present, and Future*, Shuch, H. P. (ed.), Springer.

Ehman, J. 1998. The Big Ear Wow! Signal: What We Know and Don't Know About It After 20 Years, http://www.bigear.org/wow20th.htm

Ehman, J. 2007. The Big Ear Wow! Signal: 30th Anniversary Report, http://www.bigear.org/Wow30th/wow30th.htm

Kraus, J. D. 1979. We Wait and Wonder, *Cosmic Search* 1 No. 3, 31–34.

Kraus, J. D. 1976. *Big Ear*, Cygnus-Quasar Books.

Kraus, J. D. 1966. *Radio Astronomy*, McGraw-Hill, New York.

Chapter 3. Small SETI Radio Telescope

Gray, R. H. 1985. A Small SETI Radio Telescope, *Sky & Telescope*, April, 354–356.

Gray, R. H. 1985. Small SETI Systems, in *The Search for Extraterrestrial Intelligence*. Proceedings of NRAO Workshop, May 20–22, Green Bank.

Chapter 4. META

Colomb, F. R., Hurrell, E., Lemarchand, G. A., and Olald, J. C. 1995. Results of Two Years of SETI Observations with META II, in *Progress in the Search for Extraterrestrial Life*, Shostak, G. S. (ed.), ASP Conf. Ser., Vol. 74, 345–352.

Cordes, J. M., and Lazio, T. J. 1991. Interstellar Scattering Effects on the Detection of Narrow-Band Signals, *AJ* 376, 123–134.

Drake, F. D., and Helou, G. 1978. The Optimum Frequencies for Interstellar Communications as Influenced by Minimum Bandwidths, NAIC Report 76, Ithaca, New York.

Gray, R. H. 1994. A Search of the Wow Locale for Intermittent Radio Signals, *Icarus* 112, 485–489.

Horowitz, P., and Sagan, C. 1993. Five Years of Project META: An All-Sky Narrow-Band Radio Search for Extraterrestrial Signals, *ApJ* 415, 218–235.

Horowitz, P., and Hill, W. 1989. *The Art of Electronics*, Cambridge Univ. Press.

Horowitz, P., *et al.* 1986. Ultranarrowband Searches for Extraterrestrial Intelligence with Dedicated Signal-Processing Hardware, *Icarus* 67, 525–539.

Horowitz, P. 1978. A Search for Ultra-Narrowband Signals of Extraterrestrial Origin, *Science* 201, 733–735.

Lazio, T. J. W., Tarter, J., and Backus, P. R. 2002. Megachannel Extraterrestrial Assay Candidates: No Transmissions From Intrinsically Steady Sources, *AJ* 124, 560–564.

Chapter 5. Very Large Array

Cordes, J. M., and Lazio, T. J. 1991. Interstellar Scattering Effects on the Detection of Narrow-Band Signals, *AJ* 376, 123–134.

Gray, R. H., and Marvel, K. B. 2001. A VLA Search for the Ohio State Wow, *ApJ* 546, 1171–1177.

Palmer, P. and Zuckerman, B. 1972. *The NRAO Observer* 13, No. 6, p. 26.

Perley, R. A., Schwab, F. R., and Bridle, A. H. (ed.) 1994. *Synthesis Imaging in Radio Astronomy*, San Francisco: ASP Conf. Ser. Vol. 6.

Chapter 6. SETI Down Under

Ellingsen, S. 1996. Class II Methanol Masers in Star Formation Regions, PhD Thesis, University of Tasmania.

Gray, R. H., and Ellingsen, S. 2002. A Search for Periodic Emissions at the Wow Locale, *ApJ* 578, 967–971.

Chapter 7. Other Worlds

Alfren, H. 1964. On the Formation of Celestial Bodies, *Icarus* 3, 57–62.

Audouze, J., and Israel, G. (eds.) 1994. *The Cambridge Atlas of Astronomy*, Cambridge University Press.

Bailey, C. 1926. *Letter to Herodotus* in *Epicurus: The Extant Remains*, Oxford: Clarendon Press.

Batalha, N. M., *et al.* 2012. Planetary Candidates Observed by Kepler, III: Analysis of the First 16 Months of Data, http://arxiv.org/pdf/1202.5852v1.pdf.

Boorstin, D. J. 1983. The Discoverers, Random House.

Borucki, W. J., *et al.* 2010. Kepler Planet-Detection Mission: Introduction and First Results, *Science* 327, 977.

Bruno, Giordano 1584. *On the Infinite Universe and Worlds*, translation by Dorothea Waley Singer in *Giordano Bruno: His Life and Thought*, Henry Schuman Publishers, New York, 1950, cited in *The Quest for Extraterrestrial Intelligence* by Donald Goldsmith and Tobias Owen.

Butler R. and Marcy P. 1996. A Planet Orbiting 47 Ursae Majoris, *ApJ*, 464, L153–L156.

Crowe, M. J. 1986. *The Extraterrestrial Life Debate 1750–1900: The Idea of a Plurality of Worlds from Kant to Lowell*, Cambridge University Press.

Dole, S. H. 1970. Computer Simulations of the Formation of Planetary Systems, *Icarus* 13, 494–508.

Dole, S. H. 1964. *Habitable Planets for Man*, Blaisdell, New York, http://www.rand.org/content/dam/rand/pubs/commercial_books/2007/RAND_CB179-1.pdf.

Lineweaver, C. H., and Grether, D. 2003. What Fraction of Sun-like Stars Have Planets?, *ApJ* 598, 1350–1360.

Lucretius. *De Rerum Natura* (On the Nature of the Universe), a translation by Professor Mary-Kay Gamel.

Marcy, G., Butler, R. P., Fischer, D., *et al.* 2005. Observed Properties of Exoplanets: Masses, Orbits and Metallicities, Progress of Theoretical Physics Supplement 158, 24–42.

Marcy, G., and Butler, R. P. 1996. The Planetary Companion to 70 Vir, *ApJ Letters* 464, L147.

Mayor, M., Marmier, M., Lovis, C., *et al.* 2011. The HARPS search for southern extra-solar planets XXXIV. Occurrence, mass distribution and orbital properties of super-Earths and Neptune-mass planets, preprint, *arXiv:1109.2497v1*.

Mayor, M., and Queloz, D. 1995. A Jupiter-mass companion to a solar-type star, *Nature* 378, 355–359.

Traub, W. A., 2011. Terrestrial, Habitable-Zone Exoplanet Frequency from Kepler, preprint, http://arxiv.org/abs/1109.4682.

Wolszczan, A. 1994. Confirmation of Earth-Mass Planets Orbiting the Millisecond Pulsar PSR B1257+12, *Science* 264: 5158, 538–542.

Chapter 8. Life

Billingham, J. (ed.) 1981. Life in the Universe, MIT Press.

Horowitz, N. H. 1986. *To Utopia and Back: The Search for Life in the Solar System*, W. H. Freeman.

Kasting, J. F., Whitmire, D. P., and Reynolds, R. T. 1993. Habitable Zones around Main Sequence Stars, *Icarus* 101, 108–128.

Lovelock, J. E. 1979. *Gaia: A New Look at Life on Earth*, Oxford University Press.

Matcher, J. C., and Boslough, J. 1996. *The Very First Light*, Basic Books.

McKay, D. S., *et al.* 1996. Search for Past Life on Mars: Possible Relic Biogenic Activity in Martian Meteorite ALH84001, *Science* 273, 924–930.

Miller S. L., and Urey, H. C. 1959. Organic Compound Synthesis on the Primitive Earth, *Science* 130, 245.

Nouvian, C. 2007. *The Deep: The Extraordinary Creatures of the Abyss*, University of Chicago Press, http://www.thedeepbook.org/.

Postgate, J. 1994. *The Outer Reaches of Life*, Cambridge University Press.

Ward, P. D., and Brownlee, D. 2000. *Rare Earth: Why Complex Life is Uncommon in the Universe*, Copernicus Books.

Wetherill, G. W. 1996. The Formation and Habitability of Extrasolar Planets, *Icarus* 119, 219–238.

Chapter 9. Intelligence and Technology

Alvarez, L. W., *et al.* 1980. Extraterrestrial Cause for the Cretaceous–Tertiary Extinction, *Science* 208, 4448:1095–1108.

Darwin, C. 1859. *On The Origin of Species by Means of Natural Selection: or The Preservation of Favored Races in the Struggle for Life*, John Murray, Albemarle Street, London.

Jerison, H. J. 1973. *The Evolution of Brain and Intelligence*. Academic Press, New York.

Kurzweil, R. 2005. *The Singularity is Near*, Viking, New York.

Mikkelsen, T. S. , et al. 2005. Initial Sequence of the Chimpanzee Genome and Comparison with the Human Genome, *Nature* 437, 69-87.

Moravec, H. 1988. *Mind Children*, Harvard University Press.

Newman, M. E. J., and Palmer, R. G. 1999. Models of Extinction: A Review, http://www.lassp.cornell.edu/newmme/science/ModelsOfExtinction.pdf.

Jerison, H. J. 1973. *The Evolution of Brain and Intelligence*. Academic Press, New York.

Penzias, A. A. and Wilson, R. W. (1965). A Measurement of Excess Antenna Temperature at 4080 Mc/s, ApJ 142, 419-421.

Sagan, C., *et al.* 1973. The Number of Advanced Galactic Civilizations, in *Communications With Extraterrestrial Intelligence*, Sagan, C. (ed.), 164–187, MIT Press.

Chapter 10. Interstellar Communication

Ekers, R. D., Cullers, K. C., Billingham, J., and Scheffer, L. K. (eds.) 2002. *SETI 2020: A Roadmap for the Search for Extraterrestrial Intelligence*, SETI Press, Mountain View, California.

Lightman, A. P. 1984. A Fundamental Determination of the Planetary Day and Year. *Am. J. Phys* 52 (3), 211–214.

McConnell, B. S. 2001. *Beyond Contact: A Guide to SETI and Communicating with Alien Civilizations*, O'Reilly.

Oliver, B. M. 1977. The Rationale for a Preferred Frequency Band: The Water Hole, in *The Search for Extraterrestrial Intelligence*, NASA SP-419, 63–74.

Oliver, B. M., and J. Billingham 1971. *Project Cyclops: A Design Study for a System for Detecting Extraterrestrial Life*, NASA CR 114445.

Sagan, C. (ed.) 1973. *Communication with Extraterrestrial Intelligence*, MIT Press.

Sagan, C., and Drake, F. D. 1975. The Search for Extraterrestrial Intelligence, *Sci Am* 232, 80–89.

Seeger, C. L. 1977. Notes on Search Space, in *The Search for Extraterrestrial Intelligence*, NASA SP-419, 111–125.

Sullivan, W. T. III, Brown, S., and Wetherill, C. 1978. Eavesdropping: The Radio Signature of the Earth, *Science* 199, 377–388.

Chapter 11. Search Strategies

Drake, F. D., and Sagan, C. 1973. Interstellar Radio Communication and the Frequency Selection Problem, *Nature* 245, 257–258.

Dyson, F. J. 1960. Search for Artificial Stellar Sources of Infrared Radiation, *Science* 131, 1667.

Gray, R. H. 1990. Isotropically Detectable Interstellar Beacons, *JBIS* 43, 531–536.

Gray, R. H. 1977. Broadcast Strategies in the Galaxy, *JBIS* 30, 341–343.

Gulkis, S., Olsen, E. T., and Tarter, J. 1979. A Bimodal Search Strategy for SETI, IAU SETI Conference, Montreal, Canada.

Kasting, J. 2010. The Spectroscopic Search for Life, in *How to Find a Habitable Planet*, Princeton University Press.

Kingsley, S, and Ross, M. 2011. Optical SETI: Moving toward the light, in *Searching for Extraterrestrial Intelligence: Past, Present, and Future*, Shuch, H. P. (ed.), Springer.

Lorimer, D. R., *et al.* 2007. A Bright Millisecond Radio Burst of Extragalactic Origin, Science 318, 777-779.

Maccone, C. 2011. Focusing the Galactic Internet (gravitational lens), in *Searching for Extraterrestrial Intelligence: Past, Present, and Future*, Shuch, H. P. (ed.), Springer.

Oliver, B. M. 1981. Search Strategies, in *Life in the Universe*, Billingham, J. (ed.), 364, MIT Press, Cambridge.

Oliver, B. M. 1979. The Rationale for the water hole, in *Communication with Extraterrestrial Intelligence*, Billingham J. and Pesek, R. (eds.), Pergamon.

Pfleiderer, J. 1988. Galactic Communication with Small Duty Cycles, in *Commercial Opportunities in Space*, Shahrokhi, F. *et al.* (eds.), AIAA Progress in Astronautics and Areonautics 110, 305–317.

Townes, C. 1961. At What Wavelengths Should We Search For Signals From Extraterrestrial Intelligence? *Proc Nat Acad Sci* 80, 1147.

Whitmire, D. P., and Wright, D. P. 1980. Nuclear Waste Spectrum as Evidence of Technological Extraterrestrial Civilizations, *Icarus* 42, 149–156.

Chapter 12. Contact

Billingham, J., *et al.* (eds.) 1994. *Social Implications of the Detection of an Extraterrestrial Civilization*, SETI Institute.

Carrigan, R.A. 2006. Do potential SETI signals need to be decontaminated?, *Acta Astronautica* 58, 112–117.

Declaration of Principles Concerning Activities Following the Detection of Extraterrestrial Intelligence, International Academy of Astrophysics (IAA), April 1989.

Draft Declaration of Principles Concerning the Sending of Communications to Extraterrestrial Intelligence, International Academy of Astrophysics.

Dumas, S. 2011. A Proposal for an Interstellar Rosetta Stone, in *Communication with Extraterrestrial Intelligence*, Vakoch, D. A. (ed.), SUNY Press.

Dutil, Y., and Dumas, S. 2001. Error Correction Scheme in Active Seti, American Institute of Aeronautics and Astronautics, IAA-01-IAA.9.1.09.

Freudenthal, H. 1960. *Lincos: Design of a Language for Cosmic Intercourse.* North-Holland Publishing, Amsterdam.

Harrison, A. 2011. After Contact - Then What?, in *Searching for Extraterrestrial Intelligence: Past, Present, and Future*, Shuch, H. P. (ed.), Springer.

Vakoch, D. A., 2011. Integrating Active and Passive SETI Programs, in *Communication with Extraterrestrial Intelligence*, Vakoch, D. A. (ed.), SUNY Press, Albany, New York.

Chapter 13. History

Bowyer S., *et al.* 1996. Twenty Years of SERENDIP, the Berkeley SETI Effort: Past Results and Future Plans, in *Astronomical and Biochemical Origins and the Search for Life in the Universe*, Cosmovici, C. B., Bowyer, S., and Werthimer, D. (eds.), Editrice Compositori, Bologna, p. 667.

Bowyer, S., Zeitland, G. M., Tarter, J., Lampton M., and Welch, W. J. 1983. The Berkeley Parasitic SETI program, *Icarus* 53, 147–155.

Cocconi, G., and Morrison, P. 1959. Searching for Interstellar Communication, *Nature* 183, 844–846.

Colomb, F. R., Hurrell, E., Lemarchand, G. A., and Olald, J. C. 1995. Results of Two Years of SETI Observations with META II, in *Progress in the Search for Extraterrestrial Life*, Shostak, G. S. (ed.), ASP Conf. Ser., Vol. 74, 345–352.

Davidson, K. 1999. *Carl Sagan: A Life*, Wiley & Sons, New York.

DeBoer, D. R., *et al.*, 2004. The Allen Telescope Array, Proc. SPIE Vol. 5489, Ground-based Telescopes, Oschmann, J.M. (ed.), 1021–1028.

Dixon, R. S. 1974. An Isotropic Frequency-independent Antenna System for Simultaneous Omnidirectional Doppler Shift Removal, *IEEE Trans. Antennas Propag.*, AP-22 5, 707–709.

Dixon, R. S. 2011. A Sentry on the Universe, in *Searching for Extraterrestrial Intelligence: Past, Present, and Future*, Shuch, H. P. (ed.), Springer.

Drake, F. D. 1985. Project Ozma: The Search for Extraterrestrial Intelligence, in *Proceedings of the NRAO Workshop*, Green Bank, West Virginia, May 20–22, Kellermann, K.I., and Seielstad, G.A., (eds.), p. 23.

Drake, F. D. 1961. Project Ozma, *Physics Today* 14, 140.

Ekers, R. D., Cullers, K. C., Billingham, J., and Scheffer, L. K. (eds.) 2002. *SETI 2020: A Roadmap for the Search for Extraterrestrial Intelligence*, SETI Press, Mountain View, California.

Ellingson, S. W., Hampson, G. A., and Childers, R. K. 2008. Argus: An L-Band All-Sky Astronomical Surveillance System, *IEEE Trans. Ant. Propagat.* 56, No. 2.

Ewen, H. I. 2011. Facts Concerning the Detection of the 21 cm Interstellar Hydrogen Line, by Doc Ewen, Aug. 7, 2011 (private communication).

Ewen, H. I. 1998. Interview of Dr. Harold "Doc" Ewen by K. D. Stephan on January 28, 1998 Niels Bohr Library & Archives, American Institute of Physics, http://www.aip.org/history/ohilist/6659.html.

Ewen, H. I., and Purcell, E. M. 1951. Observation of a Line in the Galactic Radio Spectrum: Radiation from Galactic Hydrogen at 1,420 Mc./sec., *Nature* 168, 356.

Gulkis, S., Olsen, E. T., and Tarter, J. 1979. A Bimodal Search Strategy for SETI, *IAU SETI Conference*, Montreal, Canada.

Jansky, K. G. 1933. Electrical Disturbances Apparently of Extraterrestrial Origin, *Proc. IRE* 21, 1387–98.

Kaplan, S. A. (ed.) 1969. *Extraterrestrial Civilizations Problems of Interstellar Communication*, NASA TT F–631.

Kardashev, N. S. 1967. Transmission of Information by Extraterrestrial Civilizations, in *Extraterrestrial Civilizations,* Tovmasyan, G. M. (ed.), NASA TT F–438, Translated from Russian, Israel Program for Scientific Translations, Jerusalem.

Korpela, E., Cobb, J., Werthimer, D., and Lebofsky, M., 2004. SETI@home Reobservation Report, http://seticlassic.ssl.berkeley.edu/newsletters/newsletter22.html

Lemarchand, G. 2005. 15 Years Developing SETI from a Developing Country, http://www.seti-argentina.com.ar/

McDonough, T. R. 1996. A Quantum Leap for SETI: Project BETA Goes On-line, *Planetary Report* 16, 4.

Murray, B., *et al.* 1978. Extraterrestrial Intelligence: An Observational Approach, *Science* 199, 485–492.

Oliver, B. M. 1985. The NASA SETI Program: An Overview, in *The Search for Extraterrestrial Intelligence*, NRAO Green Bank, 121–133.

Oliver, B. M., and J. Billingham 1971. *Project Cyclops: A Design Study for a System for Detecting Extraterrestrial Life*, NASA CR 114445.

Oliver, B. M. 1997. *The Selected Papers of Bernard M. Oliver*, Cutler, L., Fazarinc, Z., Hornak, T., and Seeger, C. (eds.), Hewlett-Packard Co.

Reber, G. 1944. Cosmic Static, *ApJ* 100, 279.

Rigden, J. S. 2002. *Hydrogen: The Essential Element*, Harvard University Press.

Sagan, C., and Shklovski, I. S. 1966. *Intelligent Life in the Universe*, Holden-Day.

Shuch, H. P. (ed.) 2011. *Searching for Extraterrestrial Intelligence: SETI Past, Present, and Future*, Springer-Verlag Praxis.

Siemion, A., *et al.* 2008. New SETI Sky Surveys for Radio Pulses, http://seti-athome.berkeley.edu/berkeley_pulse_search_paper_nov_2008.pdf

Sullivan, W. T., *et al.* 1997. A New Major SETI Project Based on Project SERENDIP Data and 100,000 Personal Computers, in *Astronomical and Biochemical Origins and the Search for Life in the Universe*, Cosmovici, C. B., Bowyer, S., and Werthimer, D. (eds.), Editrice Compositori, Bologna.

Swift, D. W. 1990. *SETI Pioneers—Scientists Talk About Their Search for Extraterrestrial Intelligence*, University of Arizona.

Tarter, J. 2011. ATA: A Cyclops for the 21st Century, in *Searching for Extraterrestrial Intelligence: Past, Present, and Future*, Shuch, H. P. (ed.), Springer.

Tarter, J. 1995. Summary of SETI Observing Programs, SETI Institute.

Troitskii, V. S. *et al.* 1979. Search for radio emissions from extraterrestrial civilizations, in *Communication with Extraterrestrial Intelligence*, Billingham J. and Pesek, R. (eds.), Pergamon Press.

van de Hulst, H. 1945. Radio Waves from Space: Origin of Radiowaves, *Nederlands tijdschrift voor natuurkunde*, 11, translated in *A Source Book in Astronomy and Astrophysics*, 1900–1975, 627–632, Lang, K. R. and Gingerich, O. (eds.), Harvard University Press, 1979.

Waldrop, M. M. 2011. SETI Is Dead, Long Live SETI, *Nature* 475, 442-444.

Welch, J., *et al.* 2009. The Allen Telescope Array: The First Widefield, Panchromatic, Snapshot Radio Camera for Radio Astronomy and SETI, *Proceedings of the IEEE*.

Chapter 14. Worries

Annis, J. 1999. Astrophysical Explanation for the Great Silence, *JBIS* 52, 19.

Comins, N. F. 1993. *What if the Moon Didn't Exist? Voyages to Earths that Might Have Been*, HarperCollins, New York.

Hart, M. H. 1975. An Explanation for the Absence of Extraterrestrial Life on Earth, *Quarterly Journal of the Royal Astronomical Society* 16, 128–135.

Jones, E. M. 1985. "Where Is Everybody?" An Account of Fermi's Question, Los Alamos National Laboratory, LA-10311-MS, UC-34b, CIC-14 Report Collection, Issued March 1985. http://www.fas.org/sgp/othergov/doe/lanl/la-10311-ms.pdf

Lada, C. J. 2006. Stellar Multiplicity and the IMF: Most Stars are Single, *ApJ* 640, L63–L66.

Lineweaver, C. H., Fenner, Y., and Gibson, B. K. 2004. The Galactic Habitable Zone and the Age Distribution of Complex Life in the Milky Way, *Science* 303 (5654): 59-62.

Oliver, B. M. 1993. Symmetry in SETI, *Third Decennial US-USSR Conference on SETI*, ASP Conference Series Vol. 47, Shostak, S. (ed.).

Sagan, C. 1983. The Solipsist Approach to Extraterrestrial Intelligence, *Quarterly Journal of the Royal Astronomical Society* 24, 113.

Tipler, F. J. 1980. Extraterrestrial Intelligent Beings Do Not Exist. *Quarterly Journal of the Royal Astronomical Society* 21, 267–281.

Ward, P. D., and Brownlee, D. 2000. *Rare Earth: Why Complex Life is Uncommon in the Universe*, Copernicus Books, New York.

Webb, S. 2002. *If the Universe Is Teeming with Aliens... Where Is Everybody? Fifty Solutions to Fermi's Paradox and the Problem of Extraterrestrial Life*, Springer.

Zuckerman, B. and Hart, M. H. (eds.) 1982. *Extraterrestrials: Where are They?*, Pergamon Press, New York.

General

First Contact: Scientific Breakthroughs in the Hunt for Life Beyond Earth, Marc Kaufman, Simon & Schuster, New York, 2011.

If the Universe Is Teeming with Aliens... Where Is Everybody? Fifty Solutions to Fermi's Paradox and the Problem of Extraterrestrial Life, Stephen Webb, Springer, 2002.

Intelligent Life in the Universe, Carl Sagan and I. S. Shklovskii, Holden-Day, San Francisco, California, 1966.

Is Anyone Out There? — *The Scientific Search for Extraterrestrial Intelligence*, Frank Drake and Dava Sobel, Delacorte Press, New York, 1992.

Life in the Universe, Jeffrey Bennett, Seth Shostak, and Bruce Jakosky, Addison Wesley, San Francisco, California, 2003.

The Quest for Extraterrestrial Intelligence, Donald Goldsmith and Tobias Owen, University Science Books, Sausalito, CA 2001.

Searching for Extraterrestrial Intelligence: SETI Past, Present, and Future, H. Paul Shuch, H. P. (ed.) Springer-Verlag Praxis, 2011.

SETI 2020: A Roadmap for the Search for Extraterrestrial Intelligence, Ronald D. Ekers, D. Kent Cullers, John Billingham, and Louis K. Scheffer (eds), SETI Press, Mountain View, California, 2002.

SETI Pioneers—Scientists Talk About Their Search for Extraterrestrial Intelligence, David W. Swift, University of Arizona Press, 1990.

Sharing the Universe: Perspectives on Extraterrestrial Life, Seth Shostak, Berkeley Hills Books, Berkeley, California, 1998.

LINKS

This section lists selected SETI and astronomical Internet sites. Some current and past searches have sites providing background, news, and other information, and many resources used by astronomers are available on the Internet such as images, catalogs, and software tools.

MORE ON THE WOW

"A Search for Periodic Emissions at the Wow Locale" by Robert H. Gray and Simon Ellingsen, *Astrophysical Journal*, 2002.
   http://www.bigear.org/Gray-Ellingsen.pdf

"A VLA Search for the Ohio State 'Wow'" by Robert H. Gray and Kevin B. Marvel, *Astrophysical Journal*, 2001.
   http://www.bigear.org/Gray-Marvel.pdf

"Interstellar Signal from the 70s Continues to Puzzle Researchers" by Seth Shostak.
   http://www.seti-inst.edu/epo/news/features/interstellar-signal-from-the-70s.php

"Still no sense in signal" by David Whitehouse.
http://news.bbc.co.uk/2/hi/science/nature/1122413.stm

"Still Searching for the Elusive 'Wow!'" by Amir Alexander.
http://planetary.org/news/2002/1021_Still_Searching_for_the_Elusive_
Wow.html

"The Big Ear Wow! Signal: What We Know and Don't Know About It After 20
Years", by Jerry R. Ehman, 1998.
http://www.bigear.org/wow20th.htm

"The Big Ear Wow! Signal (30th Anniversary Report)" by Jerry R. Ehman.
http://www.bigear.org/Wow30th/wow30th.htm

"The 'Wow!' Signal Still Eludes Detection" by Amir Alexander.
http://planetary.org/news/2001/0117_The_Wow_Signal_Still_Eludes.
html

"We Wait and Wonder" by John D. Kraus, *Cosmic Search* 1, 31–34, 1979.
http://www.bigear.org/vol1no3/wonder.htm

Wow Coordinates, original epoch B1950 and precessed to J2000.

Beam	B1950		J2000	
	RA	Dec	RA	Dec
East	19^h 22^m 22^s	-27° 03′	19^h 25^m 28^s	-26° 57′ 01″
West	19^h 25^m 12^s	-27° 03′	19^h 28^m 17^s	-26° 56′ 50″

Beam	Galactic		Ecliptic	
	L	B	Lon	Lat
East	11° 39′ 32″	-18° 54′ 17″	289.022479°	-4.910956°
West	11° 54′ 20″	-19° 29′ 28″	289.649212°	-5.000992°

Wow Frequency
1420.356 ±0.005 MHz (Kraus); 1420.4556 ±0.005 MHz (Ehman)

Wow Flux Estimates
54 Jy (Ehman); 212 Jy (Childers)

Ohio State University Radio Observatory
http://www.bigear.org/

North American AstroPhysical Observatory (Big Ear and Argus)
http://www.naapo.org/
http://www.naapo.org/Argus/data/display.html

Cosmic Search magazine electronic version
http://www.bigear.org/CSMO/HTML/CSIntro.htm

SETI PROJECTS

Allen Telescope Array
http://www.seti.org/ata
http://ral.berkeley.edu/ata/memos
http://www.setilive.org/

Amateur SETI: Project BAMBI
http://www.bambi.net/

Archives of SETI Observing Programs (historical list)
http://www.seti.org/seti/seti-background/previous-searches/
http://astrosurf.com/luxorion/Documents/seti-programs.xls

Astropulse
http://setiathome.berkeley.edu/ap_faq.php

Columbus Optical SETI Observatory
http://www.coseti.org/

Harvard University META and BETA projects
http://seti.harvard.edu/seti/harvard_seti.html

Harvard University Optical SETI projects
http://seti.harvard.edu/oseti/

Hat Creek Radio Observatory
http://www.hcro.org/

Ohio State University Radio Observatory ("Big Ear")
http://www.bigear.org/

Optical SETI at Berkeley
http://seti.ssl.berkeley.edu/opticalseti/

Princeton University Optical SETI project
http://observatory.princeton.edu/oseti/

SERENDIP (Search for Extraterrestrial Radio Emissions from Nearby
Developed Intelligent Populations)
http://seti.berkeley.edu/serendip/

SEVENDIP (Search for Extraterrestrial Visible Emissions from Nearby
Developed Intelligent Populations)
http://seti.berkeley.edu/opticalseti

SETI Argentina and META II
http://www.seti-argentina.com.ar/15-years-developing-seti-from-a-developing-country

SETI Australia Center, Southern SERENDIP
http://seti.uws.edu.au/

SETI@home
http://setiathome.berkeley.edu/

SETI Institute
http://www.seti.org

SETI Italia
http://www.seti-italia.cnr.it/

SETI League
http://www.setileague.org/

The SETI Research & Community Development Institute Limited, Amateur
SETI movement in Australia
http://www.seti.org.au/

Exoplanets

California & Carnegie Planet Search
http://exoplanets.org/

Extrasolar Planets Encyclopedia
http://exoplanet.eu/

Kepler "A Search for Habitable Planets"
http://kepler.nasa.gov/
http://kepler.nasa.gov/Mission/discoveries/candidates/
http://kepler.nasa.gov/Mission/discoveries/papers/

Astronomical Images

Aladin: Sophisticated software for accessing, displaying, comparing, and
analyzing images and catalogs from many sources and many wavelengths.
Allows users to create and annotate very good sky maps.
http://aladin.u-strasbg.fr/

DSS: The STScI Digitized Sky Survey
The National Geographic Society-Palomar Observatory Sky Survey in digital form, made with the 48-inch Schmidt telescope. Users can select targets by name or coordinates, image size, and so on. Images are returned as GIF or FITS files.
http://stdatu.stsci.edu/dss/

FIRST: Faint Images of the Radio Sky at Twenty-Centimeters
A high-resolution radio sky survey made with the VLA, covering 25% of the sky in the direction on the North and South Galactic Caps. Images contain sources down to 1 mJy (background rms noise is 0.15 mJy). "Image Cutouts" can be retrieved via Internet for any coordinates covered by the survey in either GIF for FITS file format.
http://sundog.stsci.edu/top.html

Google Sky
Images from NASA/Space Telescope Institute, Sloan Digital Sky Survey, and Digital Sky Survey Consortium, plus infrared and microwave.
http://earth.google.com/sky/

Hubble Space Telescope images
http://oposite.stsci.edu/pubinfo/Subject.html

Hubble Ultra Deep Field in 3D and HD
http://www.youtube.com/watch?v=oAVjF_7ensg

Jet Propulsion Lab images from planetary missions
http://www.jpl.nasa.gov/

NASA Photo Resources List: NASA web sites
Links to many collections including GRIN (Great Images In NASA)
http://history.nasa.gov/photo_links.html

NRAO Image Gallery
http://images.nrao.edu/

NVSS: The NRAO VLA Sky Survey (21 centimeters)
Radio sky survey made with the VLA covering 82% of the celestial sphere north of declination -40° and detecting sources down to about 2.5 mJy. A "postage stamp server" returns radio images of the sky in FITS or JPEG format. http://www.cv.nrao.edu/nvss/postage.shtml

Photopic Sky Survey
Amateur 5,000 megapixel mosaic image of the sky; fun to zoom and pan.
   http://skysurvey.org/

SkyView Virtual Observatory
SkyView retrieves images from astronomical surveys at many wavelengths and displays them or returns files. Images are requested by name (like "M31" or "Andromeda") or coordinates, with image type specified by wavelength or survey name; image size and other parameters (the galaxy M31, for example, covers three degrees). Advanced features allow positions of objects from roughly one hundred astronomical catalogs to be overlaid on images as well.
   http://skyview.gsfc.nasa.gov/skyview.html

CATALOGS

High Energy Astrophysics Science Archive Research Center (HEASARC)
Access to many astronomical catalogs. Users can search all catalogs at once for objects near given coordinates.
   http://heasarc.gsfc.nasa.gov/W3Browse/

Smithsonian Astronomical Observatory (SAO) catalog, "258,996 stars"
   http://tdc-www.harvard.edu/software/catalogs/sao.html

ORGANIZATIONS

American Astronomical Society
   http://www.aas.org/

Astronomical Society of the Pacific
http://www.astrosociety.org/

NASA Astrophysics Data System
http://adswww.harvard.edu/

NASA
http://www.nasa.gov/

National Optical Astronomy Observatory
http://www.noao.edu/

National Radio Astronomy Observatory
http://www.nrao.edu/

Planetary Society
General information on SETI, solar system research, and bioastronomy.
http://planetary.org/

SETI Institute
Supports research on life in the universe and operates the Allen Telescope
Array (with the University of California at Berkeley).
http://www.seti-inst.edu/

SETI League
International organization dedicated to privatizing the electromagnetic
Search for Extra-Terrestrial Intelligence.
http://www.setileague.org/

Sky & Telescope magazine SETI resources
http://www.skyandtelescope.com/resources/seti

Society of Amateur Radio Astronomers
http://radio-astronomy.org/

SOFTWARE

Astronomical Almanac, calculates positions of astronomical objects (longitude of site must be specified in hours—decimal longitude divided by 15 and re-expressed in *hh:mm:ss* format).
  http://www.jb.man.ac.uk/vlbi/inter/almanac.html

Drake Equation interactive calculators
  http://astro.nomical.info/drake.equation
  http://www.pbs.org/wgbh/nova/space/drake-equation.html

Doppler Shift Calculator
  http://www.narrabri.atnf.csiro.au/observing/obstools/velo.html

Excel spreadsheets for SETI and radio calculations
  http://www.setileague.org/software/spreadsh.htm

FITS File Editor
  http://heasarc.gsfc.nasa.gov/docs/software/ftools/fv/

FitsView, a FITS file viewer
Most professional astronomical images are stored in FITS-format files (Flexible Image Transport System) which require special software like FitsView to view and manipulate.
  http://www.cv.nrao.edu/~bcotton/fitsview.html

Functions and procedures for astronomy in the IDL programming language.
  http://idlastro.gsfc.nasa.gov/contents.html

SAOImage DS9, a visualization program for FITS images including many analysis tools. Can also retrieve images from some astronomical databases.
  http://hea-www.harvard.edu/RD/ds9/

SpectraPLUS, a very sophisticated audio spectrum analyzer program.
  http://www.spectraplus.com/

SPECTRAN, a spectral analysis program.
http://www.weaksignals.com/

Spectrum Lab, a sophisticated audio spectrum analyzer program.
http://www.qsl.net/dl4yhf/spectra1.html

StarGen simulates formation of planetary systems.
http://www.eldacur.com/~brons/NerdCorner/StarGen/StarGen.html

HARDWARE

"Small Radio Telescope" developed by MIT Haystack Observatory.
http://www.haystack.mit.edu/edu/undergrad/srt/index.html

"Small Radio Telescope" sold by Custom Astronomical Support Services, Inc.
http://www.cassicorp.com/

Radio gear for high frequency amateur use.
http://www.downeastmicrowave.com/

Radio telescope receivers and related products for amateur use.
http://www.radioastronomysupplies.com/

OTHER

Archives and library of National Radio Astronomy Observatory
http://www.nrao.edu/archives/
http://www.nrao.edu/library/

Dumb Or Overly Forced Astronomical Acronyms Site (or DOOFAAS)
http://www.cfa.harvard.edu/~gpetitpas/Links/Astroacro.html

Essential Radio Astronomy (ERA) is a one-semester course intended for astronomy graduate students and advanced undergraduates with backgrounds in astronomy, physics, or engineering.
http://www.cv.nrao.edu/course/astr534/ERA.shtml

European Single Dish School in the Era of Arrays: presentations on many facets of radio astronomy, from Max-Planck-Institut für Radio Astronomie.
http://www.mpifr.de/div/effelsberg/SummerSchool/

Radio Astronomy Observatories
http://www.nas.edu/bpa1/US_Radio_Astronomy_Observatories.htm
http://www.nas.edu/bpa1/NonUS_Radio_Astronomy_Observatories.htm/
http://en.wikipedia.org/wiki/List_of_radio_telescopes

SAO/NASA Astrophysics Data System: Digital Library for Physics and Astronomy (searchable library of many scientific articles).
http://adsabs.harvard.edu

FORMULAE

Antenna gain
The gain factor G of an antenna, where η is antenna aperture efficiency (typically between 0.5 to 0.7 for large reflector antennas), A is antenna area, and λ is wavelength (both in the same units, often meters), is:

$$G = (4 \, \eta \, \pi \, A) \, / \, \lambda^2$$

Beamwidth, parabolic antenna
The half-power beamwidth BW in degrees of an antenna with a diameter of D at a wavelength of λ (both in the same units, often meters) is approximately:

$$BW = 70 \, \lambda \, / \, D$$

Binomial distribution
If p is the probability that an event will occur, then in a random group of n independent trials, P_B is the probability that the event will occur x times. (from *Handbook of Space Astronomy and Astrophysics*, Second Edition, by Martin V. Zombeck, p. 410):

$$P_B\left(x,\, n,\, p\right) = \frac{n!}{X!\left(n-r\right)!}p^x\left(1-p\right)^{n-x}$$

Detection threshold

For n independent samples, the probability of error P_e, that one or more samples will exceed a value of Z_m in the absence of a real signal is (from Thompson, Moran, & Swenson 1994):

$$P_e = 1 - (1 - \exp\left(-\frac{Z_m^2}{2\sigma^2}\right))^n$$

Lyman series

Lyman spectral line wavelengths, oscillator strengths, and line strengths are shown in the table below. Line strengths were calculated using a formula from C.W. Allen (in *Astrophysical Quantities*, 1973) where S is line strength, gf is oscillator strength, and λ is wavelength (in micrometers):

$S = 32.92 \, gf \, \lambda$

Line Name	Wavelength	Wavelength	gf	Line Strength
	(angstroms)	(micrometers)		calculated
Lyman A	1215	0.1215	0.5549	2.2194
Lyman B	1025	0.1025	0.1055	0.3559
Lyman C	972	0.0972	0.0387	0.1238
Lyman D	949	0.0949	0.0186	0.0581
Lyman E	937	0.0937	0.0104	0.0321
Lyman F	930	0.0930	0.0064	0.0196
Lyman G	926	0.0926	0.0043	0.0131
Lyman H	923	0.0923	0.0029	0.0088
Lyman I	920	0.0920	0.0021	0.0064
Lyman J	919	0.0919	0.0016	0.0048

Power required

To produce a flux S at a range of R light years requires an effective isotropic radiated power P_{eirp} in watts (after Seeger, 1979, p. 109, in *Communication with Extraterrestrial Intelligence*) of:

$P_{eirp} = 1.12 \times 10^{33} \, S \, R^2$

For an isotropic broadcast, where transmitter antenna gain is 1, the power is P_{eirp} . For a broadcast from an antenna with gain G, the power P is:

$$P = P_{eirp} / G$$

Resolution

The angular resolution in radians of a telescope with diameter D at a wavelength of λ (both in the same units) is approximately:

$$R = \lambda / D$$

Sensitivity of radio telescopes

The minimum detectable flux in W/m^2, where α is the desired signal-to-noise ratio, k is Boltzmann's constant 1.38×10^{-23} Joules/K, T_{sys} is system temperature in degrees Kelvin, n is aperture efficiency, D is antenna diameter in meters, b is bandwidth in Hz, and t is integration time in seconds:

$$F_{min} = (4 \alpha k T_{sys} / \pi n D^2) (b/t)^{1/2}$$

Sun-like stars within various ranges

The number of Sun-like stars within r_0 light years is approximately (after Oliver, On the Required Sensitivity of SETI Systems, *Collected Papers*):

$$N_* = int (r_0 /8.6)^3$$

Assumes a uniform distribution of stars, which does not hold for ranges greater than 1,000 light years from the Sun due to the concentration of stars in the galactic disk.

r_0	10	50	100	500	1,000
N_*	2	200	1,600	200,000	1,600,000

Wavelength

The wavelength λ in meters corresponding to a frequency f in Hz, where c is the velocity of light (299,792,458 meters per second), is:

$$\lambda = c / f$$

FACES

John Kraus
Photo courtesy of Kraus estate.

Bob Dixon
Photo courtesy of Bob Dixon.

Jerry Ehman
Photo courtesy of Jerry Ehman.

Paul Horowitz, young radio operator.
Photo courtesy of Paul Horowitz.

Paul Horowitz, older interstellar radio builder.
Photo courtesy of Paul Horowitz.

Patrick Palmer
Photo courtesy of Patrick Palmer.

Kevin Marvel
Photo courtesy of Kevin Marvel.

Simon Ellingsen
Photo courtesy of Simon Ellingsen.

Frank Drake
Photo courtesy of Seth Shostak.

Philip Morrison
Source: NASA.

Barney Oliver
Photo courtesy of Seth Shostak.

John Billingham
Photo courtesy of Seth Shostak.

Carl Sagan
Photo courtesy of the Planetary Society.

Jill Tarter
Photo courtesy of Jill Tarter.

INDEX